Energy Management in Buildings Withdrawn

Managing the consumption and conservation of energy in buildings is the concern of both building managers and occupants, and this energy use in buildings accounts for about half of UK's energy consumption. The need to manage this has been given new emphasis by the introduction of the Climate Change Levy in April 2002.

Energy Management in Buildings introduces students and energy managers to the principles of managing and conserving energy consumption in buildings that people use for work or leisure. Energy consumption is considered for the provision of space heating, hot water supply, ventilation and air-conditioning. The author introduces the use of standard performance indicators and energy consumption yardsticks, and discusses the use and application of degree days.

This new edition includes two new chapters on current regulations and environmental impact of building services. It closely follows recent benchmarking published by CIBSE and the DEFRA energy efficiency Best Practice Programme, and covers three quarters of Unit 18 in the new HND in building services engineering. Examples in the earlier chapters of the first edition have been updated in this one, and further examples are introduced to provide a more comprehensive coverage of this important subject.

Keith Moss spent 13 years in contracting and consulting before moving to Bath where he taught to HND and Degree level. During this time he undertook private work to keep abreast of developments in the industry. In that time, he has been an external examiner and verifier for BTEC/Edexcel, and served on the CIBSE Education Training and Membership Committee.

Also available from Taylor & Francis

Energy Management in Buildings

2nd edition

Keith J. Moss

Taylor & Francis
Taylor & Francis Group

LONDON AND NEW YORK

First published 1997
by E&FN Spon

Second edition published 2006
by Taylor & Francis
2 Park Square, Milton Park, Abingdon, Oxon OX14 4RN

Simultaneously published in the USA and Canada
by Taylor & Francis
270 Madison Ave, New York, NY 10016, USA

Taylor & Francis is an imprint of the Taylor & Francis Group

© 1997, 2006 Keith J. Moss

Typeset in Sabon by
Integra Software Services Pvt. Ltd, Pondicherry, India
Printed and bound in Great Britain by
TJ International Ltd, Padstow, Cornwall

British Library Cataloguing in Publication Data
A catalogue record for this book is available from the British Library

Library of Congress Cataloging in Publication Data
A catalog record for this book has been requested

ISBN10: 0–415–35391–2 Hardback
ISBN10: 0–415–35392–0 Paperback

ISBN13: 9–78–0–415–35391–5 Hardback
ISBN13: 9–78–0–415–35392–2 Paperback

Contents

Preface

Energy Management in Buildings is a textbook for undergraduate courses in building services engineering, building engineering, environmental engineering, BTEC higher national diploma and higher national certificate in building services engineering. Since the management of energy in buildings requires an understanding of the behaviour of the building structure as well as the services to changes in outdoor climate, part of the text at least is appropriate to students studying architecture, building, facilities management and building surveying.

Following the Climate Change Levy and the Government's White Paper on energy and sustainability, there has been increased focus on the effects of fossil fuel consumption and the exploitation of finite resources. Forty-five per cent of the UK's energy consumption is taken up by its use in the provision of space heating, space cooling, lighting, communications, hot water supply and cooking.

Industry in the form of manufactured products has made considerable progress in cutting the consumption of fossil fuel and electricity, much of which is derived from fossil fuel, by changing its manufacturing processes and installing energy efficient plant.

Some of the larger organisations now have sophisticated energy conservation programmes for the services in their building stock. It is necessary now to encourage those organisations which have no energy management policy to join the fossil fuel and sustainability stewardship campaign. Negotiating a competitive fuel tariff with a fuel supplier should be considered only as the first step in the management of energy consumption on the site or campus.

This book addresses the methodologies of estimating annual energy consumption, undertaking energy audits, and monitoring and targeting energy consumption in the form of fossil fuels. It discusses the background of each chapter and this is followed up with the appropriate underpinning knowledge, and examples and case studies. References are made to source organisations, journals and articles which are pertinent to this important subject.

Acknowledgements

Grateful thanks are due to a number of part-time students at Bath College, some of whom are working in the field of facilities management, who have contributed to the preparation of the book and in teasing out many of the finer points in presentation. I am also grateful to the following organisations, some of whom have allowed me to reproduce material relevant to the subject.

> Department for Environment, Food and Rural Affairs
> The Chartered Institution of Building Services Engineers
> The Building Services Research and Information Association
> The Energy Efficiency Office (Department of Energy)
> The Met. Office
> The Building Research Energy Conservation Support Unit
> The Energy Technology Support Unit
> The Heating and Ventilating Contractors Association

Those who have assisted me in the preparation of the book have not checked the arithmetic or solutions to examples and case studies. The author holds responsibility for these.

Introduction

Energy management in buildings is a branch of the discipline of building services engineering. It is not essential that students wishing to specialise in energy management should have a prerequisite knowledge and experience of building services. In practice the background of many individuals who currently have a responsibility as energy managers or facilities managers varies widely. Many would agree, however, that underpinning knowledge about the services within the building can be of great help and continuing professional development in this area for those without it is invaluable.

This book, however, is written in such a way as to develop basic skills in energy management regardless of prior professional training, and where necessary the reader is directed to other reading matter in the series.

Each chapter of the book is set out with the nomenclature used, an introduction, worked examples and case studies, data and text appropriate to the topic and concludes with a chapter closure which identifies the skills and competences acquired. The level of mathematics needed to gain full benefit from the text is between GCSE and A level and is introduced in the solutions where necessary.

The use of computer programs and spread sheets have not been used in the presentation material since the purpose of the book is to develop the underpinning knowledge of the subject. It is left to the reader to investigate the ever moving market for software in this field.

Energy management is a moving feast; it is still a relatively young discipline and much of what is known has to be derived from historical data in the form of fuel invoices equated with building type, level of services specifications, level of building specifications and level of proactive maintenance.

Since the first edition of this book in 1997, there has been an extraordinary impetus given to frugally managing energy that is derived from fuels that when burnt produce climate changing combustion products. Furthermore, there is increasing awareness of the exponential consumption of finite resources and indiscriminate generation of unnecessary waste that is at present discarded.

It is therefore even more imperative for the energy manager to keep abreast of current developments, publications and practices.

Chapter 1

The economics of space heating plants

Nomenclature

Q	design heat load/loss (kW)
U	thermal transmittance coefficient (W/m^2 K)
A	area (m^2)
N	number of air changes per hour (h^{-1})
V	volume of space (m^3)
F_1, F_2	heat loss ratios
t_c	dry resultant or comfort temperature (°C)
t_{ao}	outdoor air temperature (°C)
t_{ai}	indoor air temperature (°C)
SDD	Standard Degree Days (K·day)
f_r	thermal response factor
Y	thermal admittance (W/m^2 K)
Q_g	indoor heat gains (kW)
t_b	Base temperature (°C)
d	temperature rise due to indoor heat gains (K)
d_t	design temperature difference (K)
t_n	minimum daily outdoor temperature (°C)
t_x	maximum daily outdoor temperature (°C)
S	number of days in the period under review
DD	Corrected Degree Days (K·day)
t_m	mean outdoor air temperature (°C)
k	Hitchin's location specific constant
e	exponent, 2.7183
C_v	ventilation conductance (W/K)
CV	calorific value MJ/litre (kg)

1.1 Introduction

The use of fossil fuels in industrialised countries has been the subject of international interest since Kioto. Much work has been done in the UK and elsewhere to reduce the consumption of energy derived from fossil fuels since that time particularly in the manufacturing industries. However, energy used in building services is estimated at 45% of national primary energy consumption in the UK.

At present 160 million buildings in the European Union use more than 40% of Europe's energy and generates a similar percentage of carbon dioxide emissions. Space heating accounts for 55% of energy consumption, domestic water heating 25%, non-domestic water heating 9% and lighting 11%.

The building services industry therefore has a mandate here to design systems which conserve energy, to provide accurate forecasts of energy consumption, to promote energy conservation, to undertake energy audits and to monitor and target the consumption of energy in buildings.

Clearly the building services engineer will be responsible for design, specification, installation and commissioning, but he or she should be able to estimate the energy consumption of a projected new or refurbished building in a professional manner and undertake an energy audit of an existing building. Facility managers are responsible for the life of the services within the building and costs of operating the services. Technical innovation in recent years has meant, for example, that boiler plant and associated equipment is more efficient in operation, bringing the benefits of lower consumption in the use of primary energy and less harmful releases of the products of combustion into the atmosphere. Pumps and fans fitted with variable speed control have an extended life and a much reduced electrical power consumption over time in most applications.

1.1.1 Towards a sustainable future

However, the need to drastically reduce carbon dioxide emissions will ultimately require a more radical approach to the way indoor space is heated in the world's temperate climates and of the life cycle costs of the heating systems and products used. A holistic view considers whole life costing that includes capital costs and operating and energy costs during the building life cycle.

For a sustainable approach to buildings and their services systems, operating cost will need to include the social and environmental costs in relation to extraction of the raw materials, the manufacturing process, the activities of construction/installation, the product life, maintenance during the product life, recycling and waste disposal.

It does now appear that controlling the release of carbon dioxide and pollutants and more respectful use of raw, finite materials is just beginning to influence decision making.

Chapters 11 and 12 consider these concepts in more detail.

1.2   The economics

Using the whole life cost model, the social and environmental costs relating to building services provision should be embedded in a sustainability appraisal that will include:

- Capital cost
- Life cycle costs
- Investment appraisal.

Investment appraisal is considered in Chapter 8.

1.2.1   Capital costs

This will include:

- Design fees
- Fees for supervising the installation of the services
- Material and labour costs of the installation
- Commissioning costs and handover
- Costs for supplying the utilities of gas, water and electricity, and
- Builders work and attendance costs.

1.2.2   Life cycle costs

It is now recognised that the costs involved during the life of the services systems within the building outweigh their initial capital cost.

Careful selection of systems and correct design procedures together with the right choice of plant and controls not necessarily based on cost alone have begun to influence discerning building owners. Another important task occurs at the project completion when correct commissioning is essential for satisfactory system performance during its life.

The features relating to plant, controls and fittings given in Table 1.1 will influence the life cycle costs. Investment for the replacement or refurbishment of plant and equipment including recycling and waste disposal will form part of life cycle costing.

The ultimate aim in sustainability is to produce products that can be completely recycled at the end of their working life unless the waste remaining is biodegradable. The life of the plant and equipment may be less than the life of the distribution pipework and radiators, for example. It may therefore be prudent to consider plant and equipment separately from distribution when accounting for replacement and refurbishment. See also Chapter 8.

Table 1.1 Factors affecting life cycle costs

Reliability	Maintainability	Safety	End of life
Degree of standardisation	Spares availability	Statutory	Recycle
Spares availability	Degree of complexity	Local specific	Waste
Disposal	Maintenance intervals		
Standby requirement	Ease of maintenance		
	Down time		
	Life		

Life cycle costs will also include:

• Fuel
• Auxiliary power for boilers, pumps, fans, temperature controls, etc.
• Planned preventive maintenance
• Performance condition monitoring
• Replacement
• Recycling
• Waste disposal
• Risk assessment
• Insurance.

On large sites the client may consider buying in the supervision, operation and maintenance of the entire building services plant together with the management of its operational costs under an outsourcing agreement referred to as Contract Energy Management.

Purchase of replacement plant can also form part of an outsourcing agreement.

1.3   Energy consumption

The estimation for energy consumption for space heating depends upon:

• Design heat loss
• Seasonal efficiency of boiler plant and systems
• Duration of occupied period and activity of the occupants
• Mode of plant operation
• Thermal inertia of the building
• Internal heat sources
• Degree Days appropriate to the season and the locality.

Internal heat sources and Standard Degree Days are discussed in Sections 1.4 and 1.5.

1.3.1 Design heat loss (Q)

This is calculated for the building and is determined from the following formulae: (It is important to note here that in the calculation of weekly, monthly or annual fuel consumption using degree days, design heat loss (Q, in Watts) is used.)

$$Q = (\Sigma(UA)F_1 + C_v \cdot F_2)(t_c - t_{ao}) \qquad (1.1)$$

where F_1 and F_2 are the heat loss factors and ventilation conductance $C_v = NV/3 \, \text{W/K}$, and $Q =$ loss through the building envelope Q_f plus loss resulting from infiltration of outdoor air Q_v.

The heat loss factors can be determined from data in the CIBSE Guide book A. The heat flow path resulting from a system of natural draught convectors will start at the indoor air point and proceed through the dry resultant and the mean radiant point to outdoor air temperature. In the case of a system of high temperature radiant strip or tube the resulting heat flow path commences at the mean radiant point. Thus the heat loss ratios are affected by the type of heating system proposed for the building and in turn the building design heat loss is also affected. This matter is discussed at length in another publication in the series.

If, however, the building is thermally insulated to current standards and has low infiltration rates which is to say that it is well sealed from ingress of outdoor air, design heat loss Q can be determined in the traditional manner from:

$$Q = (\Sigma(UA) + C_v)(t_c - t_{ao})$$

with only limited loss in accuracy.

1.3.2 Seasonal efficiency of boiler plant and systems

Boiler manufacturers quote efficiencies of their products under test conditions at full load. The efficiency of modern conventional boilers ranges from 80 to 90% compared with boilers manufactured over 15 years ago where the range was in the region of 70–80%.

There is therefore a strong argument to replace boiler plant that is over 15 years old.

A well-designed conventional boiler should maintain its efficiency to a turn down ratio of 30% or less of full load. Thereafter its efficiency may fall away and this is one of the reasons why modular boilers are recommended for plants with energy outputs above 100 kW.

Condensing boilers on the other hand have efficiencies in excess of 95% in appropriate applications. They can be used with conventional boilers to improve the overall efficiency of the plant, particularly at low load and low temperature. Seasonal efficiency differs from efficiency under test conditions since it accounts for variations in load throughout the heating season.

In the case of the generation of hot water supply, central storage heated indirectly from the boiler plant will have a lower seasonal efficiency than direct fired instantaneous HWS generation. Table 1.2 provides a guide to seasonal efficiencies.

1.3.3   Duration of the occupied period

This varies with the use to which the building is put. For buildings intermittently occupied, like offices for example, space heating plant can be shut down at night and over weekends. It is normal to provide frost protection where plant will be triggered if indoor temperature falls below around 12 °C.

It is important that the building envelope on the warm side of the thermal insulation layer does not fall too low in temperature, to avoid a long preheat period in addition to its original purpose of providing protection from water systems freezing. Consideration needs to be given to how the cleaning staff are accommodated when time scheduling the plant.

For buildings intermittently occupied corrections are made to the annual SDD total. This matter is addressed in Chapter 3.

Buildings continuously occupied are continuously heated during the heating season. Space heating plant can have the facility of night setback that will cause a dip in the thermal capacity of the building envelope as well as the designed drop in indoor temperature.

Table 1.2 Suggested seasonal efficiencies for plant and system

Type of system	Seasonal efficiency (%)
Continuous space heating	
Condensing/conventional boilers, compensated system	85
Fully controlled oil-gas-fired conventional boiler serving radiator/convector system	70
As above with multiple modular boilers and sequence control	75
Intermittent space heating	
Condensing/conventional boilers, compensated system	80
Fully controlled oil-gas-fired conventional boiler serving radiator/convector system	65
As above with multiple modular boilers and sequence control	70
Domestic hot water	
Gas-oil-fired conventional boiler and central storage	60
Direct gas-oil-fired instantaneous water heaters	75

1.3.4 Effects of mode of plant operation on energy consumption

Plant energy output is dependent upon the mode of plant operation during the heating season and the thermal inertia of the building envelope.

For continuously heated buildings having a high thermal inertia (this corresponds to a building envelope consisting of high density materials), plant energy output will not have to respond to the full downward swings in outdoor temperature during the heating season unless it is sustained over many days. This is due to the thermal flywheel effect of the building envelope which is also assisted by high density constituents of the internal walls and floors.

For space heating plant operated intermittently, plant energy output does not vary much with variations in the thermal inertia of the building envelope.

For buildings with a low thermal response factor and intermittently occupied, the preheat period is relatively short and so is the cooldown period after plant shutdown that will take place just before the building is vacated.

On the other hand, buildings having a high thermal inertia will require a longer preheat before occupation but the plant can be shut down some time before the building is vacated without loss in thermal comfort.

The effect on the length of the preheat period will, however, be significant after a weekend shutdown for a building envelope having a high thermal inertia.

Location of the thermal insulation in the building envelope will also have an effect upon the length of the preheat period. Insulation located on the inside surface of external walls and ceilings of an envelope which otherwise has a high thermal inertia will effectively change the characteristics of the envelope to one of low thermal inertia – that is to say, there will be a rapid response to the switching of the heating plant. This is because there is little thermal mass on the warm side of the insulation and in consequence the preheat and cool down periods will be short.

It should be emphasised here that the design heat load used for estimating annual energy consumption must not include a plant margin (overload capacity or boosted plant output). Allowances for the daily, weekly and annual occupation times for buildings heated intermittently are made by correcting the SDD for the locality. This is investigated in Chapter 3.

1.3.5 Effects of mode of plant operation on the thermal characteristics of the building envelope

As space heating plant becomes more intermittent in operation in response to occupancy patterns, heat flow absorbed by the building envelope takes on significance from heat flow transmitted through the envelope. This is because the external structure of the building is cold and will absorb larger quantities of heat energy than that required for transmission through the structure.

The CIBSE Guide publishes tables of Thermal Transmittance Coefficients (U values) and Admittance (Y absorption values). For a cavity wall with 50 mm of glass fibre slab on the warm side of the air cavity, the U value is $0.46 \, \text{W/m}^2 \, \text{K}$ and the Y value is $3.7 \, \text{W/m}^2 \, \text{K}$. This clearly demonstrates that the absorption of heat energy by the cold external wall is significantly greater than the transmission of heat energy through it – eight times more significant in fact.

Once that part of the external wall on the warm side of the thermal insulation has risen from its datum temperature (t_d) following a shutdown period to its optimum temperature, at the end of the preheat period, occupation of the building can take place without complaints of discomfort which otherwise would result from excessive body heat loss by heat radiation to the cold internal surfaces of the external envelope. It might be argued that highly intermittent occupation patterns therefore require the design heat loss to be determined from Admittance Y in place of Transmittance U. In practice it is recommended that space heating plant is operated for a time each day in buildings with highly intermittent occupation patterns during the winter season to maintain the temperature of the building envelope. It follows therefore that the plant is sized as for buildings intermittently occupied.

This approach avoids the deterioration of decoration and fabrics and the potential risk of condensation.

The net result in terms of annual energy consumption is that similar buildings with intermittent and highly intermittent occupation patterns will have comparable energy consumption levels. Three examples of buildings likely to have highly intermittent occupancy patterns are churches, village halls and club houses.

1.4   Estimation of indoor heat gains (Q_g)

Indoor heat gains (Q_g) are not normally considered when determining the plant energy output unless they are continuous. The automatic temperature controls on the space heating plant should account for internal heat gains. They should also account for solar heat gains, particularly through glazing, which occur on some winter days when the sun is at low altitude.

Indoor heat gains are, however, considered when estimating weekly, monthly or annual fuel consumption. It is therefore necessary to access heat output data from various sources such as artificial lighting, computers, photocopiers, cooking and laundry equipment, display units, etc. The BSRIA Rules of Thumb can be used as resource material although it is important not to overestimate the internal heat gains. For example, heat gains from modern desktop computers and low energy light fittings are much reduced when compared with similar equipment of 20 years ago. Knowledge of internal heat gains allows the determination of the Base temperature (t_b) which is defined as the outdoor temperature above which no heating is required since design indoor temperature will be maintained by the internal heat gains.

The temperature rise (d) due to the effect of internal heat gains is determined from:

$$d = Q_g/(\Sigma(UA) + C_v)\,\mathrm{K}$$

Alternatively $\quad d = Q_g/(Q/d_t)\,\mathrm{K}$

where d_t = design temperature difference between indoors and outdoors.
Thus Base temperature $t_b = t_c - d\,°\mathrm{C}$

1.4.1 Determination of Base temperature, t_b

Base temperature for space heating is that temperature outdoors at and above which heating is not required. With heating Degree Days the lower the Base temperature (indicating higher indoor heat gains) the lower will be the Degree Days and hence the lower the energy consumption.

Internal heat sources directly affect the determination of Degree Days. Standard Degree Days are calculated from various Base temperatures as indicated in the CIBSE Guide book A.

The Base temperature used in this book is taken as 15.5 °C.

Example 1.1
The design heat loss for a building is 210 kW for indoor and outdoor design temperatures of 20 °C and −3 °C respectively. From the data given below, calculate the temperature rise due to indoor heat gains and hence determine the Base temperature.

Data
Heating effect from artificial lighting 10 W/m²
Heating effect from office equipment 12 W/m²
Total floor area 2300 m²

Solution
From the data, indoor heat gains $Q_g = (10 + 12) \times 2300 = 50\,600\,\mathrm{W}$
Indoor temperature rise $d = 50.6/(210/23) = 5.54\,\mathrm{K}$
Base temperature indoors $t_b = t_c - d = 20 - 5.54 = 14.46\,°\mathrm{C}$
Base temperature = 14.46 °C

Conclusion
- Clearly since Base temperature varies from the common value of 15.5 °C a correction must be applied to the weekly, monthly and annual SDD.

See Table 1.4. CIBSE Guide book A lists heating degree days for W. Pennines (Manchester) for Base temperatures of 10, 12, 14, 15.5, 16, 18, 18.5 and 20 °C. Appendix 1 adopts the Base temperature of 15.5 °C to calculate Standard Degree Days.

- Base temperature is also known as Balance temperature, which is related to continuous performance monitoring and is discussed in Chapter 10.
- Historically, indoor heat gains from lighting and from office equipment such as computers and photocopiers have been overestimated and in any event high efficiency lighting systems and modern computers emit substantially less heat than earlier equipment. For this reason it is sensible to exercise caution. The BSRIA Rules of Thumb guide is a useful reference (Appendix 8).
- The inclusion of heat gains from the occupants is not considered as it has the effect of lowering the Base temperature and hence the corrected annual number of Degree Days. Heat gains need to be continuous and at a consistent level before being included in the calculation of the rise in indoor temperature resulting from heat gains.

1.5   Standard Degree Days

The determination of Degree Days is based upon daily maximum (t_x) and minimum (t_n) outdoor temperatures and the temperature rise (d) resulting from internal heat gains. Fuel Efficiency booklet No. 7 is published by the Energy Efficiency Office in which there is a map of the UK identifying the locations from which weather data is collected for the calculation of SDD. The map shows Degree Day Isopleths. Isopleths are isograms which are lines drawn on the map connecting points having equal numbers of Degree Days.

Standard Degree Days for 17 regions in the UK and for a 20-year period up to 1979 are given in the CIBSE Guide book B (1986) and reproduced in Table 1.5. This table gives SDD totals for the months of September to May – a 9-month heating season.

The current CIBSE Guide book A (1999) tabulates the annual SDD for 18 regions for the 20-year period up to 1995 to a Base temperature of 15.5 °C (Appendix 1). It accounts for the 12 months of a year so it is important to subtract the months when no heating is normally available, usually June, July, August, September and probably October.

The bimonthly edition of DEFRA's Energy and Environmental Management gives the preceding monthly SDD for 18 regions together with the 20-year averages.

Monthly SDD totals for a specified locality vary from corresponding months of each year due to variations in the seasons in temperate climates. Annual SDD totals therefore vary from year to year as one would expect. Have a look at Appendix 1 and Table 1.5 in which two separate 20-year periods are listed. Do not forget that the table in Appendix 1 gives 12-month totals whereas Table 1.5 gives 9-month totals. If the annual SDD totals for the 20-year periods ending in 1979 and 1995 are compared over the September to May period there is a strong indication that the UK climate is warming up.

1.5.1 Local climates and micro-climates

The existence of local climatic conditions that vary from the 18 regional locations that the Met. Office uses to generate SDD is well known. Micro-climates are found in local climates and can differ again. This can occur in sheltered and exposed valleys and on hill sites or near rivers and lakes as well as in densely built-up areas. Instead of using the Met. Office data, owners of building complexes can have their own weather stations to generate monthly and annual SDD via dedicated software. Either way the facilities manager can log, compare and monitor current climatic conditions with site energy consumption on a monthly basis.

1.5.2 Standard Degree Days – Limitations and a definition

Standard Degree Days are dependant upon outdoor climate and indoor heat sources other than the space heating plant. They do not account for solar heat gain in winter or wind speed which adds a chill factor to the maximum and minimum outdoor temperatures if they are taken in a typically ventilated and screened local meteorological station.

Within these limitations the annual and monthly totals of SDD therefore provide the means of comparing over different periods and in different geographical locations the variations in load sustained by heating plants. They also can be used to check the consistency or otherwise of the performance of a heating plant on a monthly or annual basis.

Example 1.2
A community heating scheme to a housing estate operates in the Thames Valley area. Determine the increase in energy consumption for a similar scheme projected for a location in Midlands.

Data
Take the actual annual number of SDD as 2034 for the Thames Valley and 2357 for the Midlands (Table 1.5).

It is clearly apparent that the severity of the climate directly affects the number of Degree Days recorded for a region.

Solution
The SDDs given in this example are taken from Table 1.5.

Estimate of the energy increase $= (2357 - 2034)/2034 = 0.159 = 15.9\%$

There now follows two examples in which monthly energy consumption is checked against the SDD for that month.

Example 1.3
A building uses 150 litres of heating oil during a winter month having 380 DD. The consumption in the previous month having the same number of DD was 144 litres. Calculate the apparent loss on plant efficiency.

Solution
Apparent loss in efficiency $= 100 \times (150 - 144)/144 = 4.16\%$

Example 1.4
A building energy manager logs a fuel consumption of 178 litres of oil during a winter month having 341 DD. According to the records of a previous month having 351 DD, the log shows an oil consumption of 170 litres. Determine the apparent effect if any on plant efficiency.

Solution
Current fuel consumption $= 178/341 = 0.522$ litres/DD
Previous fuel consumption $= 170/351 = 0.484$ litres/DD
Apparent loss in efficiency $= 100 \times (0.522 - 0.484)/0.484 = 7.85\%$

Conclusions
- A plot of monthly energy consumption against local DD can be made.
- The calculations do not identify the cause for the apparent loss in efficiency in Examples 1.3 and 1.4.
- The apparent loss in efficiency in solutions to Examples 1.3 and 1.4 may be due to a variety of causes: the timing of a servicing contract, errors in recording fuel consumption, adjustment to the time scheduling of a heating circuit, windows left open in part of the building during redecoration, lowering of combustion efficiency in the boiler plant, thermal insulation breakdown in an external duct, etc.
- One of the tasks of the facilities manager is to find the cause.

1.6   Calculation of Standard Degree Days

The actual number of DD for a given location are assessed using maximum t_x and minimum t_n daily outdoor temperatures rather than the arithmetic mean daily temperature t_{md} outdoors.

Determination of SDD adopting the empirical Met. Office formulae of 1928.

For t_x above 15.5 °C but by a lesser amount than t_n is below

$$SDD/day = 0.5(15.5 - t_n) - 0.25(t_x - 15.5)$$

For t_x above $15.5\,°C$ by a greater amount than t_n below

$$SDD/day = 0.25(15.5 - t_n)$$

Clearly when t_x and t_n are both below $15.5\,°C$

$$SDD/day = 15.5 - 0.5(t_x + t_n)$$

Thus $SDD/day = 15.5 - t_m$

The maximum possible number of DD annually is determined from:

$$MDD = S(t_c - d - t_{ao}) = S(t_b - t_{ao})$$

where S = number of days in the heating season and is 273 from September to May; and t_{ao} = outdoor design temperature, °C.

Example 1.5
Determine the number of SDD in the sample week taken from data recorded in a specified locality. What is the maximum number of DD for the same period, given an outdoor design temperature of $-1\,°C$

Data

Day	$t_x(°C)$	$t_n(°C)$
Monday	16	8
Tuesday	18	14
Wednesday	17	8
Thursday	13	6
Friday	10	2

Solution
The solution is given in Table 1.3.

Table 1.3 Solution to Example 1.5

Day	Calculation	SDD/day
Monday	$0.5(15.5 - 8) - 0.25(16 - 15.5)$	3.625
Tuesday	$0.25(15.5 - 14)$	0.375
Wednesday	$0.5(15.5 - 8) - 0.25(17 - 15.5)$	3.375
Thursday	$15.5 - 19/2$	6.000
Friday	$15.5 - 12/2$	9.500

From Table 1.3, the actual number of SDD can be obtained by addition:

Actual number of SDD from Monday to Friday $= 22.875$

The maximum number of SDD is obtained by substituting the outdoor design temperature t_{ao} of $-1\,°C$ for a base temperature of $15.5\,°C$

Maximum number of SDD $= 5(15.5 + 1) = 82.500$

Conclusions
- The number of SDD/day increases with the severity of the outdoor climate. The number of SDD from Monday to Friday in this sample is low relative to the maximum possible number for 5 days which occurs when the outdoor design temperature remains continuously at $-1\,°C$.
- The maximum possible number of DD is calculated here so that a comparison with the recorded total SDD can be made. The building heat loss will be $22.875/82.5 = 27.7\%$ of the design heat loss from the building during those 5 days.
- Little energy is required from the heating plant during the week and then only on Thursday and Friday.
- An approximate method for calculation of SDD for each day may be undertaken by subtracting the 24-hour mean daily temperature t_m, that is the mean of t_x and t_n, from Base temperature t_b. The total for the 5 days considered in Example 1.5 on this alternate basis comes to 21.5 SDD in comparison with 22.875 SDD. You should now confirm that this is so.

This approximate method is used to determine monthly DD totals from the adoption of Hitchin's formula. See Example 1.7

1.7   Standard Degree Days to different Base temperatures

Of course many buildings will have a Base temperature different from the standard of $15.5\,°C$ (Example 1.1) that forms the basis of the SDD data in this publication. In such cases this requires a correction to be made to the published SDD totals before they can be used in the determination of estimated energy use, comparisons of energy use, or performance checks. The correction factors are given in Table 1.4. Annual SDD are published by the Meteorological Office and updated on a monthly basis. Table 1.5 below is taken from the CIBSE 1986 Guide and is the 9-month annual average for the 20-year period up to May 1979, from 1 September to 31 May.

Table 1.4 Corrections for Base temperatures other than 15.5 °C

Base temperature	DD/SDD
10	0.33
12	0.57
14	0.82
15	0.94
15.5	1.0
16	1.06
17	1.18
18	1.3

Table 1.5 Nine-month annual SDD totals for the 20-year period up to 1979, Base temperature 15.5 °C

Degree Day regions	Location	SDD
Thames Valley	Heathrow	2034
South Eastern	Gatwick	2275
Southern	Bournemouth	2130
South Western	Plymouth	1840
Severn Valley	Bristol	2109
Midlands	Birmingham	2357
West Pennines	Manchester	2233
North Western	Carlisle	2355
Borders	Boulmer	2464
North Eastern	Leeming	2354
East Pennines	Finningley	2243
East Anglia	Honington	2304
West Scotland	Glasgow	2399
East Scotland	Leuchars	2496
North-east Scotland	Aberdeen	2617
Wales	Aberport	2094
Northern Ireland	Belfast	2330

Twelve month annual SDD data for the 20-year period up to 1995 is found in Appendix 1.

The annual SDD in each region varies. This is why a 20-year period is used for the purposes of estimating annual energy consumption. A particularly cold or warm winter will therefore show different amounts of energy used for that particular year. There are therefore two uses for SDD here:

i   Estimation of projected annual energy consumption for a building or site.
ii   Comparison of actual monthly energy consumption for the site with the monthly SDD which can be undertaken each month of the winter. See Examples 1.3 and 1.4.

Monthly and annual SDD data are based upon a heating season from 1 September to 31 May which is 39 weeks or 273 days (Table 1.5). In practice, heating plant is not started until it is sufficiently cold which may extend well into October in which case the annual SDD should be adjusted accordingly.

For practical purposes the annual number of SDD therefore will vary depending upon the severity of the climate and the length of the heating season.

Example 1.6
The building discussed in Example 1.1 has a temperature rise due to indoor heat gains (d) of 5.54 K resulting in a Base temperature t_b of 14.46 °C. Determine the annual number of DD if it is located in the west of Wales.

Solution
From Table 1.5, SDD = 2094
From Table 1.4, DD/SDD = 0.82 when $t_b = 14$ °C and 0.94 when $t_b = 15$ °C
Adopting a linear interpolation for when $t_b = 14.46$ °C, DD/SDD = 0.875
Thus the corrected annual DD for this building assuming continuous heating will be DD = 2094 × 0.875 = 1832
Annual DD = 1832.

Conclusions
Note the conditions under which Table 1.5 is constituted. For example, one of the qualifications is that the SDD listed are for a heating season of 273 days starting on 1 September. If the heating system is normally started up on 1 October, the monthly average for September should be deducted before correction.

Further corrections to SDD are required for intermittent heating – see Chapter 3.

Example 1.7
A residential home located in Manchester has a design heat load of 130 kW when indoor temperature is held at 23 °C and design outdoor temperature is −4 °C. Heat gains are estimated to give a temperature rise of 4 K. The heating system consists of gas-fired conventional modular boilers and condenser boiler with weather compensated system. The cost of natural gas is 2.5 p/kWh. Estimate the annual fuel cost.

Solution
It is assumed here that the home is continuously heated.

From Table 1.5, SDD for Manchester = 2233 annually
From Table 1.4, for a Base temperature of $t_b = 23 - 4 = 18$ °C and the correction is 1.3

So annual corrected DD $= 2233 \times 1.3 = 2903$

For the period 1976–1995 (Appendix 1), the annual 9-month SDD $= 2331 - 149 = 2182$ for comparison

So the later 20-year period yields annual corrected DD $= 2182 \times 1.3 = 2837$

Now from Chapter 2, AED $= Q \times 24DD/d_t = 130 \times 24$
$$\times\, 2903/(23+4) = 335\,458\,\text{kWh}$$
$$\text{and AEC} = \text{AED/seasonal efficiency}$$
$$= 335\,458/0.85 = 394\,656\,\text{kWh}$$

where seasonal efficiency is taken from Table 1.2

Finally annual fuel cost $\text{AF}_c = 394\,656 \times 0.025 = \text{£}9866$

Example 1.8

A continuously heated workshop located near Heathrow, London, has a design heat load of 210 kW for indoor and outdoor conditions of 18 and $-2\,°C$. The heat gains in the workshop are estimated to give a temperature rise of 6 K. Seasonal efficiency of the oil-fired heating system is estimated at 70%. Estimate the annual consumption of light grade fuel oil.

Solution

Annual SDD for Heathrow from Table 1.5 is 2034. Base temperature for the workshop is $18 - 6 = 12\,°C$ and from Table 1.4, the correction on the SDD is 0.57.

Thus annual corrected DD $= 2034 \times 0.57 = 1159$

From Appendix 1 this compares, when June, July and August are subtracted from the annual total, with 1122 DD corrected as above.

Now following the procedure in Example 1.7

$$\text{AED} = Q \times 24\,\text{DD}/d_t$$
$$= 210 \times 24 \times 1159/(18+2) = 292\,068\,\text{kWh}$$
$$\text{and AEC} = 292\,068/0.7 = 417\,240\,\text{kWh}$$
$$= (417\,240 \times 3.6)\,\text{MJ}$$

Note 1 kWh = 3.6 MJ

Taking the calorific value of light fuel oil as 40.5 MJ/litre from Table 2.1, then from Chapter 2 AFC $=$ AEC/CV $= (417\,240 \times 3.6)/40.5 = 37\,088$ litres

Qualifying remarks

It should be noted that these examples provide *estimates* of cost or consumption.

1.8   Research into Standard Degree Days

An alternative method of determining SDD to any Base temperature is to use an empirical formula developed by Hitchin (1983):

$$\text{Average DD/day} = (t_b - t_m)/(1 - e^{-k(t_b - t_m)})$$

where t_m is the mean outdoor air temperature in the month obtained from daily maximum and minimum temperatures and k which varies slightly with the location has been determined from 20 years of weather data from 1952 to 1971 and has a mean value of 0.71.

The average SDD/day obtained from Hitchin's formula are then multiplied by the number of days in the month to obtain the monthly DD for a locality.

There now follows an example adopting Hitchin's formula.

Example 1.9

Given that the Base temperature for a building is calculated from indoor heat gains to be 12 °C and the average outdoor temperature for the month of November was 6 °C, determine the average SDD/day and hence the SDD for that particular month.

Solution

Substituting the data into Hitchin's empirical formula:

$$\text{Average DD/day} = (12 - 6)/(1 - e^{-0.71(12-6)})$$
$$= 6/(1 - 2.7183^{-4.26})$$
$$= 6/0.9859$$
$$= 6.09$$

The DD for the month of November will be $6.09 \times 30 = 183$ for a Base temperature of 12 °C

Conclusions

This solution is in fact not much different from SDD $= S(t_b - t_m) = 30(12 - 6) = 180$. Hitchin's formula can be used to generate monthly DD totals for use in monitoring and targeting. See Chapter 10.

1.9   Limitations of the SDD method of estimating annual energy consumption

You are reminded of the limitations of SDD:

- Daily maximum and minimum outdoor temperatures are recorded in meteorological stations in 18 locations around the UK. They do not account for other locations where micro-climates can exist and may render substantial variations to those in the nearest meteorological station.

- The determination of SDD does not account for the effects in the building of low altitude solar radiation during the winter season.
- The effects of wind chill factors are not accounted for.
- The subjective difference between a dry cold climate and a damp cold climate at a similar dry bulb temperature is not accounted for.
- SDD offered by the Met. Office are taken in 18 regions of the UK; they may not represent the local climate.

1.10 Chapter closure

This chapter has focused upon the factors that might affect the cost in use of space heating plants in temperate climates. It introduces issues relating to life cycle costing, carbon dioxide emissions and sustainability that are discussed further in Chapters 11 and 12.

Standard Degree Days have been investigated as a means by which annual energy consumption might be measured and checks in the actual performance of plant and systems can be verified on a monthly basis. The use of different Base temperatures has been introduced. Limitations on the accuracy of the application of SDD have been identified so that value judgements can be made by the practicing engineer or facilities manager.

Chapter 2

Estimating energy consumption – Continuous space heating

Nomenclature

AED   annual energy demand (kWh)
Q design heat loss (kW)
U thermal transmittance coefficient (W/m^2 K)
A area (m^2)
N number of air changes/hour
t_{ai} indoor air temperature (°C)
t_{ao} outdoor air temperature (°C)
t_c dry resultant temperature (°C)
SDD   Standard Degree Days (K · days)
DD   Corrected Degree Days (K · days)
AEC   annual energy consumption (kWh)
AFC   annual fuel consumption (litres, tonnes, m^3)
AFc   annual fuel cost
CV   calorific value (MJ/litre, kg, m^3)
P_1, P_2 proportions of heat flow
R_t total thermal resistance (m^{2}K/W)
R_{si} inside thermal resistance (m^{2}K/W)
R_{so} outside thermal resistance (m^{2}K/W)
R thermal resistance of material (m^{2}K/W)
CPV   cumulative present value
r interest rate
n length of term in years
Q_v heat loss due to infiltration of outdoor air (W)
d_t design temperature difference
C_v ventilation conductance (W/K)
V volume of space (m^3)

2.1 Introduction

This chapter focuses upon the projected energy consumption of space heating continuously heated buildings and energy savings resulting from the application of thermal insulation to existing continuously heated buildings. Projected annual energy consumption may be a requirement in the client's brief for the building services together with the determination of estimates of annual carbon dioxide and oxides of nitrogen emissions from services plant.

Recourse to sources such as the Building Regulations and CIBSE Guide book F relating to benchmarking will be a useful start. However, there is sufficient information in the Tables and Appendices here for the topics covered in this publication.

2.2   Estimating procedures for continuously heated buildings

Examples of continuously heated buildings are hospitals, clinics, residential homes, nursing homes, workshops, airports and factories on three shift operation. There are two ways of calculating the annual energy demand (AED) for a projected building namely:

- Direct use of annual DD and the thermal transmittance coefficient. This tool is useful for estimating energy savings when thermally upgrading the building envelope.
- Use of annual DD, the equivalent hours of plant operation at full load and the design heat load Q. This tool allows the calculation of annual energy estimates for existing and proposed buildings.

It has to be stressed at the outset that these procedures can only be estimates because of a number of unknown factors.

2.2.1   Continuous heating

- Level of plant and system maintenance
- Level of supervision for day-to-day plant operation
- Level of occupants' awareness to energy conservation.

2.2.2   Intermittent heating (see Chapter 3)

In addition to the factors listed for continuous heating:

- Overtime working
- Time for cleaning the premises.

These and other factors are not likely to be known at the design and specification stage of a project and will influence the AED estimate.

Building owners/occupiers can make significant savings at minimal cost by being proactive in ensuring that issues like maintenance, supervision and occupant awareness are addressed.

2.2.3 Formulae used

$$AED = 24(SDD) \times UA \, kWh \qquad (2.1)$$

Checking units of terms: $h/day \times K \cdot Days \times kW/m^2\, K \times m^2 = kWh$

$$AED = Q \times (24 \, SDD/d_t) \, kWh \qquad (2.2)$$

Checking units of terms: $kW \times h/day \times K \cdot Days/K = kWh$

$$AEC = AED/(\text{seasonal efficiency}) \, kWh \qquad (2.3)$$

Seasonal efficiency of a plant takes into account the efficiency over the heating season. Refer to Table 1.2.

$$AFC = 3.6(AEC)/CV \, \text{in litres, kg or m}^3 \qquad (2.4)$$

Note $1 \, kWh = 3.6 \, MJ$

$$AFc = AFC \times \text{cost/litre, kg or m}^3 \qquad (2.5)$$

Note in the case of natural gas which is charged in pence/kWh

$$AFc = AEC \times \text{cost/kWh} \qquad (2.6)$$

The estimation of AED for a projected plant commences with the design heat loss Q which includes the structural heat loss and that due to infiltration of outdoor air. It may be calculated in the traditional manner from dry resultant (comfort) temperature, outdoor air temperature and the heat loss factors F_1 and F_2. See Equation 1.1.

If the building envelope is well insulated, the simpler formula for design heat loss (Q, in watts) may be adopted without introducing more than a 3% error:

$$Q = (\Sigma(UA) + C_v)(t_c - t_{ao})$$

where ventilation conductance $C_v = NV/3 \, W/K$

The following example investigates the addition of thermal insulation to a roof and adopts SDD and the thermal transmittance coefficient U.

Example 2.1

The roof of an industrial building located in East Anglia is to be insulated such that its thermal transmittance coefficient is reduced from 3.13 to 0.4 W/m² K. If the roof area is 400 m², estimate the annual saving in energy demand for a continuously heated building using SDD.

Solution

Annual energy demand to offset the heat loss through the roof can be calculated from Equation 2.1. Note that thermal transmittance U has the units $kW/m^2\,K$.

You will also note that indoor/outdoor design temperatures are not given in the question and in fact are not needed for the solution. However, by adopting SDD implies use of the Base temperature of 15.5°C and continuous heating. Corrected DD can be used instead here for different Base temperatures and for intermittent heating. Refer to Chapter 3. The SDD for East Anglia is taken from Table 1.5.

From Equation 2.1 the saving in energy will be

$$AED = 24(SDD) \times (U_1 - U_2)A \, kWh$$
$$= 24 \times 2304 \times (0.00313 - 0.0004) \times 400$$
$$= 60\,383\,kWh.$$

Conclusion

Saving in energy annually $= 60\,383\,kWh$

Note:

- AED does not account for seasonal efficiency of the heating plant but this can easily be factored into the solution. See Equation 2.3.
- If the cost of fuel and seasonal plant efficiency are accounted for, the cost saving can be calculated and knowing the projected cost of the roof upgrade a simple payback period can be determined.

The following case study extends the work done in Example 2.1 by looking at cost savings using simple payback and discounted payback and accounting for seasonal efficiency of the heating plant.

Case study 2.1

The roof of a workshop in continuous operation is "flat" and consists of 12 mm stone chips, three layers of felt on 20 mm of shuttering ply over 100 mm wood joists with 1 mm of polythene and 6 mm of plaster board on the underside. It is intended to line the underside of the plasterboard with 25 mm of polyurethene board.

a   Given that the roof dimensions are 20 m × 15 m, determine the annual fuel cost estimate before and after refurbishment.
b   If the installation cost is £2500 determine:

 i   the simple payback period
 ii the payback period using a discount rate of 4%.

Data
Adopt the SDD for the Thames Valley, fuel for space heating is light grade oil at a cost of 22 p/litre, seasonal efficiency is taken as 65%.

Solution
The first part of the solution requires a knowledge of how the thermal transmittance coefficient or U value is determined. Since it is a joisted roof there is differential heat flow, namely that through the joists U_j and that through the air space between them U_a. See Figure 2.1. This calls for the adoption of the formula for the non-standard thermal transmittance coefficient $U_n = U_j P_1 + U_a P_2$ where P_1 and P_2 are the proportions of heat flow through the joists and through the air space between, taken as 30 and 70%, respectively.

The thermal conductivity of the constituent parts of the roof are taken from data in the CIBSE Guide and total thermal resistance R_t, before refurbishment, through the joisted portion of the roof is:

$$R_t = R_{so} + R_{stone\ chips} + R_{polythene} + R_{ply} + R_{joist} + R_{plaster\ board} + R_{si}$$
$$= 0.04 + 0.0125 + 0.018 + 0.1429 + 0.7143 + 0.0375 + 0.1$$
$$= 1.0652 \, \text{m}^2\text{K/W}$$

from which $U_j = 0.939 \, \text{W/m}^2 \, \text{K}$.

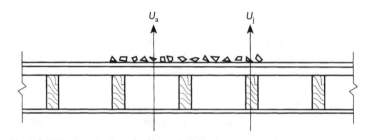

Figure 2.1 Roof detail for Case study 2.1.

Total thermal resistance through the air space between the joists R_t before refurbishment will be:

$$R_t = 0.04 + 0.0125 + 0.018 + 0.1429 + 0.16 + 0.0375 + 0.1$$
$$= 0.5109 \, m^2 \, K/W$$

from which $U_a = 1.96 \, W/m^2 \, K$.

Thus, before refurbishment $U_j = 0.939$ and $U_a = 1.96 \, W/m^2 \, K$.

After refurbishment the thermal resistance of the polyurethene insulation must be added to each calculation, from which $U_j = 0.484$ and $U_a = 0.660 \, W/m^2 \, K$.

The non-standard thermal transmittance coefficients U_n can now be determined thus:

Before refurbishment $U_n = 0.939 \times 0.3 + 1.96 \times 0.7 = 1.6537 = 0.0016537 \, kW/m^2 \, K$.

After refurbishment $U_n = 0.484 \times 0.3 + 0.66 \times 0.7 = 0.6072 = 0.0006072 \, kW/m^2 \, K$.

We are now in a position to determine the annual energy cost of heat flow through the roof of the workshop before and after refurbishment.

So before refurbishment, Equation 2.1 yields AED = 24(SDD) $\times U_n \times A$ kWh.

Heating SDD for the Thames Valley is taken from Appendix 1 for the months of October to May, giving an annual figure of 1888.

(a)
Before the improvement

$$AED = (24 \times 1888 \times 0.0016537 \times 20 \times 15) \, kWh$$

$$= 22\,480 \, kWh$$

$$AEC = 22\,480/0.65 = 34\,584 \, kWh$$

$$AFC = 3.6 \times 34\,584/40.5 = 3074 \text{ litres of light grade fuel oil}$$

$$AFc = 3074 \times 0.22 = £676$$

Before improvement AFc = £676.

Note that the calorific value of light grade fuel oil from Table 2.1 is 40.5 MJ/litre.

After the improvement
The calculation can be done by using the ratio of before and after thermal transmittance coefficients:

$$AFc = 676 \times (0.0006072/0.0016537) = £248$$

After improvement $AFc = £248$.
There is a significant difference as you can see between the estimated costs before and after refurbishment.

The second part of the question refers to both simple and discounted payback periods. The payback period is of importance since the services engineer or facilities manager will be required to produce evidence that the energy-saving measure is cost-effective. This will depend upon the attitude that the client or senior management has towards seeing a return on capital spent.

Short payback periods are unfortunately the present trend partly due to current philosophy in the marketplace. This attitude may be forced into change by the effects of dwindling resources of primary energy and the greenhouse effect, allowing periods of payback to be extended over a number of years.

If there are no costs in use of the energy-saving measure, simple payback is the cost of the measure divided by the annual savings in fuel costs thus:

(b)

(i) *Simple payback* = (cost of measure)/(net annual savings)

$$= 2500/(676 - 248)$$

Simple payback = 5.84 years

Summary part (a) and (b(i))
Cost before £676, cost after £248 and simple payback 5.84 years.

Conclusions of (a) and (b(i))
- This means that after 5.84 years the building owner will be able to reap the benefit of the saving in fuel costs of £428 per annum, the cost of the improvement having been paid for by the annual savings in that payback period.
- It is assumed here that the work will be done when the workshop is not in use otherwise loss in production must be included in the cost and this would extend the period of payback.
- Simple payback assumes that at the end of the payback period relative costs will be the same as they were at the beginning.
- Discounted payback attempts to allow for inflation and is discussed in Chapter 8.

(b)
(ii)   *Payback period at 4%*
The formula which applies here is that for cumulative present value (CPV):

$$CPV = (1 - (1 + r)^{-n})/r$$

Statistical tables are published giving values of CPV and n for ascending values of r and are included in Appendix 6.

Furthermore CPV = (cost of measure/net annual savings) = simple payback = 5.84 years, r = 4% and n is the number of years of payback. Substituting these values:

$$5.84 = (1 - (1.04)^{-n})/0.04$$

$$0.2336 = 1 - (1.04)^{-n}$$

$$(1.04)^{-n} = 0.7664$$

$$(1.04)^{n} = 1.3048$$

$$n \times \log 1.04 = \log 1.3048$$

$$n = 0.1155/0.0170$$

therefore

$$n = 6.8 \text{years}$$

Summary part (b(ii))
Discounted payback at 4% is 6.8 years.
    The higher the discount rate the longer will be the period of payback.
    For more details of financial appraisal refer to Chapter 8.

2.3   Adoption of equivalent hours of operation at full load

The term $(24 \text{ SDD}/d_t)$ is the equivalent hours of boiler plant operation *at full load* (EH). Thus $EH = (24 \text{ SDD}/d_t)$.

    There now follows an example that distinguishes between equivalent hours of plant operation at full load and total operating hours for a boiler plant.

Example 2.2

A continuously heated building located in Finningley in the East Pennines has an indoor design temperature of $20\,°C$ and an outdoor design of $-5\,°C$. Find the equivalent hours of operation at full load and the total operating hours of the boiler plant.

Solution

From Table 1.5, SDD = 2243. Equivalent hours of plant operation at full load

$$EH = 24\ SDD/d_t$$
$$= 24 \times 2243/(20 + 5)$$
$$= 2153\ hours$$

Maximum number of operating hours occurs with the plant operating continuously over the heating season of 273 days from 1 September to 31 May and

$$Maximum\ annual\ hours = 273 \times 24 = 6552\ hours.$$

Conclusion

You can clearly see the effect that the annual total of SDD has on the number of hours the plant is operating at *full* load. This of course is quite different from the total number of operating hours for the boiler plant over the winter season of 273 days. The boiler plant here is working at full load for $(2153/6552) \times 100 = 33\%$ of the total operating time. It reflects the effect of the local winter climate and the fact that outdoor temperature in the heating season is rarely as low as outdoor design temperature.

   The following example estimates the AED in kWh for a heated building using SDD and EH.

Example 2.3

A projected building having a design heat loss of 180 kW is to be located in the Midlands. Determine the annual energy demand for continuous heating during the heating season from 1 September to 31 May. Indoor and outdoor design temperatures are 19 and $-3\,°C$ respectively.

Solution

Adopting Equation 2.2 and SDD from Table 1.5,

$$AED = 180 \times \{(24 \times 2357)/(19 + 3)\} = 462\,829\ kWh$$

Note: The seasonal efficiency is not accounted for in AED. Table 2.1 lists the calorific values of different fuels.

   The example that follows estimates the site storage volume of fuel oil using SDD, EH at full load and Annual Fuel Consumption.

Example 2.4

A building located in the north-east of Scotland is to be continuously heated and has a heat loss of 220 kW at design conditions of 20 and $-5\,°C$.

Table 2.1 Calorific values of fossil fuels

Fuel	Calorific value
Gas oil 35s	37.8 MJ/litre
Light oil 250s	40.5 MJ/litre
Medium oil 1000s	40.9 MJ/litre
Butane	28.3 MJ/litre
Propane	25.3 MJ/litre
Natural Gas	38.7 MJ/m³
Coal (average)	27.4 MJ/kg
Electricity	3.6 MJ/kWh

Determine the storage volume for medium grade oil required on site for a three-week period based upon the annual SDD data taken from Table 1.5, calorific value of the oil taken from Table 2.1, with seasonal efficiency taken as 70%.

Solution
(a) From Equations 2.2, 2.3 and 2.4

$$\text{AED} = 220 \times (24 \times 2617/(20 + 5)) = 552\,710\,\text{kWh}$$

$$\text{AEC} = (552\,710)/0.7 = 789\,586\,\text{kWh}$$

$$\text{AFC} = (3.6 \times 789\,586)/40.9 = 69\,499\ \text{litres}$$

For a three-week period of the 39-week heating season, based upon SDD, the estimated volume to be stored on site will be:

$$69\,499 \times 3/39 = 5031\ \text{litres}$$

Volume of oil requiring storage $= 5346$ litres

There now follows an example of the effects of reducing the natural infiltration rate.

Example 2.5
A continually heated workshop measuring $18\,\text{m} \times 12\,\text{m} \times 4\,\text{m}$ and located in Northern Ireland is to have draught inhibiting doors fitted and it is estimated that this will reduce the average heat loss due to infiltration from 2 air changes per hour to $0.75\,\text{h}^{-1}$. Determine the annual savings estimate of medium grade fuel oil used in the space heating plant. Seasonal efficiency may be taken as 65%.

Solution
Clearly it is not a straightforward matter to estimate the reduction in infiltration of outdoor air resulting from the introduction of the new doors.

The BSRIA, however, can undertake "on site" determination on a "before and after" basis.

The ventilation conductance C_v is required here and $C_v = NV/3\,\text{W/K}$
Substituting: $C_v = (2 \times 18 \times 12 \times 4)/3 = 576\,\text{W/K} = 0.576\,\text{kW/K}$
Altering Equation 2.2 to suit, $\text{AED} = C_v \times (24\,\text{SDD})\,\text{kWh}$
Checking the units of terms $\text{AED} = \text{kW/K} \times \text{h/day} \times \text{K} \cdot \text{days}$

SDD for Northern Ireland is taken from Appendix 1 and calorific value CV is taken from Table 2.1

Substituting

$$\text{AED} = 0.576 \times 24 \times 2360 = 32\,625\,\text{kWh}$$

From Equation 2.3

$$\text{AEC} = 32\,625/0.65 = 50\,192\,\text{kWh}$$

From Equation 2.4

$$\text{AFC} = 3.6 \times 50\,192/40.9 = 4418\,\text{litres}$$

Consumption before improvement = 4418 litres
Consumption after improvement
Revised ventilation conductance $C_v = (0.75 \times 18 \times 12 \times 4)/3 = 216\,\text{W/K} = 0.216\,\text{kW/K}$
Substituting as above

$$\text{AED} = 0.216 \times 24 \times 2360 = 12\,234\,\text{kWh}$$

$$\text{AEC} = 12\,234/0.65 = 18\,822\,\text{kWh}$$

$$\text{AFC} = (3.6 \times 18\,822)/40.9 = 1657\,\text{litres}$$

A quick solution after improvement is to multiply 4418 litres by the ratio of ventilation conductances, thus $4418(0.216/0.576) = 1657$ litres
The estimate of the annual saving in fuel oil will be $(4418 - 1657) = 2761$ litres.

Example 2.6
A continuously heated residential building is to be located in the Borders and has a design heat load of 180 kW. Using the system of SDD calculate the estimated fuel consumption annually by adopting equivalent hours at full load (EH).

Data
The fuel is coal, seasonal efficiency 65%, indoor design temperature is 20 °C and outdoor design temperature is −3 °C.

Solution
Remember EH represents the actual running hours of the heating plant at full load.
From Appendix 1

   SDD = 2483 and includes June, July and August

From Equation 2.2

   $AED = Q \times (24\ SDD/d_t)\ kWh$

   $AED = 180 \times (24 \times 2483)/(20 + 3) = 466\,372\ kWh$

   $AEC = AED/\text{seasonal efficiency} = 466\,372/0.65 = 717\,496\ kWh$

From Table 2.1 the average CV for coal is 27.4 MJ/kg

   $AFC = 3.6 \times AEC/CV$

   $AFC = (3.6 \times 717\,496)/27.4 = 94\,270\ kg = 94.27\ \text{tonnes}$

Note: 1 kWh = 3.6 MJ
Estimated fuel consumption = 94 tonnes/year.

2.4   Qualifying remarks

It is most important when submitting a report of annual energy costs or savings to management or a client that it contains statements qualifying the estimate.
   There are a number of factors relating to the accuracy of the Degree Day method for the calculation of energy costs:

- Standard Degree Days do not account for wind chill.
- Solar heat gains can penetrate deeply into the building when the sun is at low altitude during the winter months contributing to internal heat gains. This is not accounted for in the determination of SDD.
- SDD are generated using data taken from Met. Office weather stations located in 18 regions of the UK. A region covers a large area and cannot reflect the effect of local climates.

- In the determination of design heat loss, Q, it is assumed that natural infiltration of outdoor air occurs simultaneously in all the rooms of a building. In fact it only takes place on the windward side. Thus the actual heat loss attributable to natural infiltration will be about 50% of the total.

2.5 Chapter closure

In this chapter we have investigated two methodologies for the determination of annual fuel estimates for continuously heated buildings. They include calculation of annual energy consumption/cost and savings for whole buildings and annual energy consumption/cost for heat flow through part of the building envelope.

You are now able to estimate the annual fuel consumption/cost for continuously heated buildings in the UK and elsewhere in temperate climates if Standard Degree Days are available. If an annual fuel cost estimate is required it is important that the best tariff is obtained from the fuel supplier, not forgetting the standing charge which may apply. This information may need to accompany the estimate when submission is made to the client.

Chapter 3

Intermittent space heating

Nomenclature

AED	annual energy demand (kWh)
AEC	annual energy consumption (kWh)
SDD	Standard Degree Days (K·days)
Q_g	heat gains indoors (kW)
Q	design heat loss (kW)
d_t	design indoor/outdoor temperature difference (K)
t_{ao}	outdoor air temperature (°C)
t_b	Base temperature (°C)
t_c	indoor comfort temperature (°C)
d	temperature rise due to indoor heat gains (K)
DD	corrected Degree Days (K·days)
AFC	annual fuel consumption (litres, kg, m³)
CV	calorific value (MJ/litre, kg, m³)
AFc	annual fuel cost (£)
HWS	hot water supply
Y	consumption of hot water (litres/person)
N	number of occupants
z	heat losses from hot water supply system (%)
E	number of days of occupation in the week
F	number of weeks of occupation in the year
PI	performance indicator (kWh/m²)
HFI	heat flux indicator (W/m²)

3.1 Introduction

Clearly not all buildings are heated continuously throughout the winter, primarily since they are not continuously occupied.

As discussed in Chapter 1, this has an effect upon the building envelope and internal structure by causing it to absorb heat energy during the preheat period and beyond, and to reject heat energy, preferably into the building, when the plant is shut down. The use of solid partitioning such as concrete blocks and floors assists in damping down this diurnal swing in the temperature within the building. The location of the thermal insulation in the building envelope also dictates the magnitude and rate of swing in the mean indoor temperature during the off-period.

These matters may be considered at the feasibility stage of a project since the pattern of occupancy can have a bearing on the location of the thermal insulation within the building envelope.

3.2   The estimation of annual energy demand

The estimation of annual energy demand, AED, for heated buildings is based upon the design heat load without the addition of a plant margin (overload capacity/boosted plant output), this is the case for intermittently heated as well as continuously heated buildings.

However, the thermal capacity and therefore the thermal response of intermittently heated buildings are accounted for, in respect of the building classification – light, medium or heavy weight (Table 3.3) – and the occupancy pattern (Tables 3.1 and 3.2). Lightweight buildings are classified as single storey factory type with little or no solid partitions or top floors of multistorey buildings when undivided.

Medium weight buildings are classified as single storey of masonry or concrete with solid partitions in concrete block.

Heavyweight buildings are those of more than one storey and constructed from masonry or concrete with solid floors and some solid partitioning.

Very heavy buildings are classified as curtain walling, masonry or concrete, especially multistorey, much subdivided within by solid partitions.

Table 3.1 Correction for the length of the occupied week

Occupied period (days)	Light building	Heavy building
7	1.0	1.0
5	0.75	0.85

Table 3.2 Correction for length of occupied day

Occupied period (hours)	Light building	Heavy building
4	0.68	0.96
8	1.00	1.00
12	1.25	1.02
16	1.40	1.03

Table 3.3 Correction for response of building and plant

Type of heating	Light building	Medium building	Heavy building
Continuous	1.0	1.0	1.0
Intermittent – responsive plant	0.55	0.7	0.85
Intermittent – plant with long time lag	0.70	0.85	0.95

Corrections are made to the weekly, monthly and annual SDD to account for these factors. These are taken from the CIBSE Guide and given in Tables 3.1, 3.2 and 3.3, respectively. Note that the location of the thermal insulation within the building envelope may change its thermal response; a heavyweight building can be turned into a building which then behaves more like a lightweight building for space heating, if during refurbishment, for example, thermal insulation is applied to the inside surface in the form of dry lining.

The indoor temperature will fall more quickly on plant shutdown. It will also rise more rapidly during the preheat period prior to occupation.

There are therefore a total of four potential corrections to annual SDD when temperature rise due to indoor heat gains is included. This is accounted for in Table 1.4 with respect to variations in Base temperature.

The corrected SDD (DD) then allows the estimation of AED for intermittently heated buildings.

3.3   The estimation of annual fuel consumption for offices

There are four categories of office in benchmarking. They are:

- Type 1: Naturally ventilated cellular
- Type 2: Naturally ventilated open plan
- Type 3: Air-conditioned standard
- Type 4: Air-conditioned prestige.

Types 1 and 2 are considered here.

The following case study considers an office heated intermittently.

Case study 3.1
Estimate the annual fuel consumption and performance indicators for a four-storey office block, Type 1, classified as a heavyweight building and located in Fife, East Scotland. The estimated heat gains from lighting and office equipment is $25\,W/m^2$.

Data

Design heat loss	115 kW
Treated floor area	360 m²/floor
Indoor design temperature	20 °C
Outdoor design temperature	−3 °C
Occupancy pattern	5 days a week, 8 hours a day
Plant operation	responsive/intermittent
Fuel	medium grade oil
Seasonal efficiency	65%

Solution

From Table 1.2 the annual 9-month SDD in Leuchars is 2496.

From Appendix 1, annual SDD is 2351 for comparison following deduction of the months of June, July and August. You will notice that the climate is slightly warmer in the 20-year period up to 1995.

$$\text{Temperature rise due to heat gains indoors } d = Q_g/(Q/d_t)$$

$$= (360 \times 4 \times 0.025)/$$

$$(115/(20+3))$$

$$= 7.2 \,\text{K}$$

$$\text{Base temperature } t_b = t_c - d = 20 - 7.2$$

$$= 12.8 \,°\text{C}$$

From Table 1.1, the correction factor for the temperature rise due to indoor heat gains, by interpolation, DD/SDD = 0.67

From Table 3.1, for 5 days a week occupation correction = 0.85

From Table 3.2, for 8 hours a day occupation correction = 1.00

From Table 3.3, for intermittent/responsive plant correction = 0.85

The corrected SDD = 2351 × 0.67 × 0.85 × 1.00 × 0.85 = 1138 DD

From Equation 2.2

$$\text{AED} = \text{design load} \times (24\,\text{DD}/d_t)\,\text{kWh}$$

$$= 115 \times (24 \times 1138/23) = 136\,560\,\text{kWh}$$

From Equation 2.3

$$\text{AEC} = 136\,560/0.65 = 210\,092\,\text{kWh}$$

From Equation 2.4

$$\text{AFC} = 3.6\,\text{AED}/(\text{CV})\,\text{MJ}$$

$$= (3.6 \times 210\,092)/(40.9)$$

Note: CV of medium grade fuel oil, from Table 2.1, is 40.9 MJ/litre.

AFC = 18 492 litres

From the above,

AEC = 136 560/0.65 = 210 092 kWh

Building performance indicator,

$$PI = AEC/(floor\ area)\ kWh$$
$$= 210 092/(4 \times 360)$$
$$= 146\ kWh/m^2$$

Building heat flux indicator (HFI) related to the design heat loss in W/m^2 is another useful guide for comparison and HFI = (design heat load)/(floor area)

$$HFI = 115 000/(4 \times 360) = 80\ W/m^2$$

Summary of Case study 3.1

$$AFC = 18 492\ litres, \quad PI = 146\ kWh/m^2, \quad HFI = 80\ W/m^2$$

Benchmarks from Appendix 4 are Good Practice 79 kWh/m² and Typical 151 kWh/m² indicating that the building and space heating system needs improvement.

The performance indicator for the building was based on treated floor area that excludes plant rooms and unheated areas. Some energy consumption benchmarks in CIBSE Guide book F (Appendix 4) are calculated on gross floor area.

Comparison with a similar building in another region
A comparison with a similar building and plant in another region of the UK to that of Case study 3.1 would be appropriate here.

Similar building and plant is located near Heathrow airport. Show a comparison with that located in Fife.

Solution
The 9-month annual SDD in Fife is 2351 whereas at Heathrow for the same 20-year period (Appendix 1) it is 1941.

The annual fuel consumption, therefore, at Heathrow will be:

$$AFC = 18 492 \times 1941/2351 = 15 267\ litres$$

The percentage reduction in annual fuel consumption for space heating will be equivalent to $(2351 - 1941)/2351 = 17.4\%$.

Case study 3.2
The following data relates to a six-storey office block located in Manchester. The building is of steel frame construction with reinforced concrete floors and infill panels of rendered concrete blockwork. It is going to be refurbished to current thermal insulation and air tightness standards as a naturally ventilated open plan office (Type 2) and re-let as office accommodation. The estimated heat gains from lighting and office equipment after refurbishment will be in the region of $20\,W/m^2$. The remaining data relates to the refurbished offices.

Design heat load	130 kW
Treated floor area	$370\,m^2$/floor
Indoor design temperature	20 °C
Outdoor design temperature	−3 °C
Occupancy	5 days/week, 12 hours/day
Plant operation	responsive/intermittent
Fuel	natural gas
Seasonal efficiency	70%
Population density	one person/10 m^2 floor
Hot water consumption	1.5 litres/person

Hot water consumption is via space heating plant using plate heat exchangers. The office is closed for 4 weeks a year.

The following solutions will be found

i   Determine the design heat flux indicator (HFI) in W/m^2 of floor.
ii Estimate the annual fuel consumption in m^3 of gas for space heating and hot water supply.
iii Determine the performance indicator in kWh/m^2 for the refurbished office and compare with the benchmarks.
iv Determine the carbon dioxide emission in kg/m^2 and compare with the benchmarks.

Solutions
i   Design heat flux indicator

HFI = design heat load/floor area

$$= 130\,000/(6 \times 370) = 58.6\,W/m^2 \text{ floor}$$

ii *Space heating*
   Standard Degree Days for Ringway, Manchester (Appendix 1) is 2228
   annually. Omitting June, July, August and September when no heating
   is required reduces the SDD to 1998.
      Corrections to SDD:

   Temperature rise due to indoor heat gains

   $$d = Q_g/(Q/d_t) = (0.02 \times 6 \times 370)/(130/23) = 7.86\,\text{K}$$

   Base Temperature $t_b = (t_i - d) = 20 - 7.86 = 12.14\,°\text{C}$
   From Table 1.4 correction = 0.6
   Corrections from Tables 3.1, 3.2 and 3.3 are 0.85, 1.02 and 0.85

Applying all four corrections to the annual SDD of 1998:

   Corrected SDD = $1998 \times 0.6 \times 0.85 \times 1.02 \times 0.85 = 883$ DD
   From Equation 2.2 AED = $130 \times 24 \times 883/23 = 119\,781\,\text{kWh}$
   From Equation 2.3 AEC = $119\,781/0.7 = 171\,116\,\text{kWh}$
   From Equation 2.4 AFC = $(3.6 \times 171\,116)/38.7 = 15\,917\,\text{m}^3$

Note the calorific value of natural gas from Table 2.1 is $38.7\,\text{MJ/m}^3$

Hot water supply
The reader should refer to Chapter 4, Equation 4.1.
From equation 4.1 AED = $(0.231\,YN(1.1)EF)/3.6\,\text{kWh}$
The number of occupants = $(6 \times 370)/10 = 222$
AED = $(0.231 \times 1.5 \times 222 \times 1.1 \times 5 \times 48)/3.6 = 5641\,\text{kWh}$
AEC = AED/(seasonal efficiency) = $5641/0.7 = 8059\,\text{kWh}$
AFC = $(3.6 \times \text{AEC})/\text{CV} = (3.6 \times 8059)/38.7 = 750\,\text{m}^3$
AEC for hot water supply is considered a Base Load as it is likely to be
   fairly constant all year round. Refer to Figure 7.3.
Total gas consumption = $15\,917 + 750 = 16\,667\,\text{m}^3$
iii The performance indicator for space heating PI = $171\,116/(6 \times 370) =$
   $77\,\text{kWh/m}^2$
   The benchmarks for a Type 2 office are 79–$151\,\text{kWh/m}^2$
iv Taking the conversion factor for natural gas from Table 6.1 as
   $0.19\,\text{kgCO}_2/\text{kWh}$, the carbon dioxide emission benchmark CEI = $77 \times$
   $0.19 = 14.63\,\text{kg/m}^2$.
      This compares with the benchmarks for the carbon dioxide emission
   index (CEI) of between 16 and $31\,\text{kg/m}^2$.
      The annual carbon dioxide emission is estimated at $14.63 \times (6 \times$
   $370) = 32\,479\,\text{kg}$. This is equivalent to 33 tonnes/annum of carbon
   dioxide for the refurbished offices and does not include the hot water
   supply.

Case study 3.3

The data for a new four-storey office located in Oxford is detailed below. Estimate the temperature rise due to indoor heat gains, the annual consumption of medium grade fuel oil for space heating and site storage for 4 weeks.

Data

Design heat load	56 kW
Estimated heat gains from lighting	$10\,W/m^2$
Estimated heat gains from computers	$5\,W/m^2$
Treated floor area	$200\,m^2$/floor
Indoor design temperature	19 °C
Outdoor design temperature	−2 °C
Occupancy pattern	5 days/week, 12 hours/day
Plant operation	responsive/intermittent
Type of building	lightweight
Seasonal efficiency	70%

Solution

Indoor heat gains

Temperature rise, $d = Q_g/(Q/d_t) = \{(10+5) \times 200 \times 4\}/(56\,000/21) =$ 4.5 K Annual energy consumption.

Annual SDD for the Thames Valley (Appendix 1) less the four summer and early autumn months $= 2033 - (49 + 20 + 23 + 53) = 1888$.

Base temperature, $t_b = (t_i - d) = 19 - 4.5 = 14.5\,°C$
From Table 1.4 the correction $= 0.88$
From Tables 3.1, 3.2 and 3.3 further corrections are 0.75, 1.25 and 0.55
The corrected SDD $= 1888 \times 0.88 \times 0.75 \times 1.25 \times 0.55 = 857$ DD
From Equation 2.2   AED $= Q \times 24$ DD$/d_t = (56 \times 24 \times 857)/21 =$ 54 848 kWh
From Equation 2.3   AEC $=$ AED/(seasonal efficiency) $= 54\,848/0.7 =$ 78 354 kWh
From Equation 2.4   AFC $= (3.6\,$AEC$)/$CV $= (3.6 \times 78\,354)/40.9 =$ 6897 litres.
Note the CV for medium grade fuel oil from Table 2.1 is 40.9 MJ/litre

Site storage of fuel oil

There are 39 weeks from September to May inclusive. The heating season taken for this case study is October to May giving 35 weeks. Site storage of oil will therefore be $6897 \times (4/35) = 788$ litres.

Summary for Case study 3.3

Temperature rise resulting from indoor heat gains 4.5 °C, annual fuel consumption 6897 litres and site storage of oil for 4 weeks 788 litres.

3.4   The estimation of annual fuel costs for a school

Case study 3.4

A secondary day school, largely single storey, located in Northern Ireland, has a design heat loss of 320 kW for an average indoor temperature of 19 °C when outdoor temperature is −1.5 °C. It is equipped with indoor sports facilities which do not include a swimming pool. The two badminton courts and an assembly hall are hired out for two nights a week and on Saturdays throughout the heating season to local community groups. The design heat loss for these areas is 22 and 27 kW respectively.

Estimate the annual energy costs for the school and the estimated annual charge for heating the out-of-hours facilities.

The space heating plant consists of three conventional boilers operating in parallel with sequence control. Seasonal efficiency is taken as 70%.

Data

Floor area of the school is 2900 m²

Fuel supply to the school is natural gas at a cost of 2 p/kWh

School holidays: Autumn half term, 1 week; Christmas, 2 weeks; spring half term, 1 week; Easter, 2 weeks; spring bank holiday, 1 week.

During term times occupancy is 5 days a week, 7 hours a day

During the holiday periods the plant is held on setback during weekdays at an indoor temperature of 15 °C except for the office accommodation the temperature controls of which can be set as required.

Solution

Note: This solution will not account for fuel consumption for the catering facilities at the school.

The 9-month SDD for Northern Ireland is 2169 from Appendix 1.

The correction factors from Tables 3.1, 3.2 and 3.3 are selected/interpolated as 0.75, 0.92 and 0.7 respectively. You will see that an intermittent plant with a long time lag has been chosen from Table 3.3 in the absence of appropriate information.

There is no reference to indoor heat gains in the question. One source of heat gain which is likely here across the whole school will be that due to lighting. Other sources such as that from computer equipment will be discounted since they are not likely to be evenly spread across the school.

A figure of 10 W/m² is a conservative estimate for a 500 lux level of illuminance from an enclosed surface mounting fluorescent luminaire (Appendix 7).

The temperature rise due to the heat gain from lighting

$$d = Q_g/(Q/d_t)$$
$$= (2900 \times 10)/(320\,000/20.5) = 1.86\,\text{K}$$

If the temperature rise is taken as 2 K resulting from the lighting, Base temperature $t_b = 19 - 2 = 17\,°C$ and from Table 1.4 the correction DD/SDD = 1.18

Applying these corrections to the SDD total for Northern Ireland:

Corrected SDD $= 2169 \times 0.75 \times 0.92 \times 0.7 \times 1.18 = 1236$ DD

From Equation 2.2

$$\text{AED} = \text{design load in kW} \times (24\ \text{DD}/d_t)\ \text{kWh}$$
$$= 320 \times 24 \times 1236/20.5 = 463\,048\ \text{kWh}$$
$$\text{AEC} = \text{AED}/(\text{seasonal efficiency})$$
$$= 463\,048/0.7 = 661\,497\ \text{kWh}$$

Building performance indicator

$$\text{PI} = \text{AEC}/(\text{floor area})$$
$$= 661\,497/2900 = 228\ \text{kWh/m}^2$$

From Equation 2.5 AFc $= 661\,497 \times 2/100 = £13\,230$

3.5 Qualifying remarks

The following qualifying remarks need to be made without which the AFc estimate is open to serious criticism.

- No account has been made for the penetration of low altitude solar radiation. Degree Days do not account for solar heat gains which will occur during the winter months.
- Holiday periods when the plant is setback during the day is not accounted for here nor is the use of the offices during the holidays. It is assumed that the one will offset the other.
- Account must now be made for the school facilities offered to the local community. The application of DD is not appropriate here. Evening occupation occurs when outdoor temperature is likely to be dipping although the use of the assembly hall on Saturdays might be more typical of a DD calculation. 39 weeks' use during the heating season at, say, 3 hours for each of two weekday evenings and 12 hours' use on Saturdays will provide the total occupancy time:

$$(39 \times 2 \times 3) + (39 \times 12) = 702\ \text{hours}$$

This is the time that the space heating plant will be required to maintain indoor design temperature. It does not represent the time the plant is operating at full load and therefore a correction factor must be applied to account for occasions when outdoor temperature is above the design value of $-1.5\,°C$.

The maximum number of Degree Days annually (MDD) for the 9-month heating season based upon design indoor and outdoor temperature for the locality will be:

MDD $= 39 \times 7(19 + 1.5) = 5597$

The SDD for Belfast $= 2169$

The correction factor which will be applied here $=$ SDD/MDD $=$ $2169/5597 = 0.388$

Thus equivalent hours of plant operation at full load, EH $= 702 \times 0.388 = 272$ hours

The design heat load $= (22 + 27)\,kW = 49\,kW$

Therefore AED $= 49 \times 272 = 13\,328\,kWh$

AEC $=$ AED/(seasonal efficiency) $= 13\,328/0.7 = 19\,040\,kWh$

Thus AFc $= 19040 \times 2/100 = £381$

Estimated annual charge for heating the out-of-hours facilities $=$ £381

Note: Another way of determining the charge is to look at the annual kWh totals.

The AEC for the school is 661 497 kWh and that for the out-of-hours facilities is 19 040 kWh. This represents 2.88% of the total annual energy consumption, and 2.88% of the annual fuel cost of £ 13 230 is £ 381.

- What further qualifying remarks should be made relating to these AFc's?
- The Good Practice and Typical benchmarks for secondary schools from Appendix 4 are $108\,kWh/m^2$ and $144\,kWh/m^2$ which indicates that the school is well above the upper limits for new design.
- Hot water supply has not been considered for the school. You should now attempt to estimate the annual energy consumption and cost based upon 500 pupils and with reference to Chapter 4.

3.6   The estimation of annual fuel consumption for a factory

The following case study relates to a factory operating a two-shift system.

Case study 3.5

A pre-1995 single-storey factory situated on a trading estate near Birmingham having a two-shift system of production has a design heat loss of

240 kW when held at 17 °C with an outdoor temperature of -3 °C. It is heated with direct gas-fired unit heaters for 6 days a week, 50 weeks a year. The factory floor area is 2000 m². Estimate the annual fuel consumption for space heating and compare the performance indicator with the benchmarks.

Solution

A production timetable of two shifts can be taken as two shifts each of 8 hours a day.

The SDD for Birmingham, from Appendix 1, is 2228.

Deducting the months of June, July, August and September, SDD is 1998.

From Tables 3.1, 3.2 and 3.3, the correction factors by interpolation where appropriate are: 0.875, 1.4 and 0.55 respectively. You should now confirm these corrections.

The temperature rise indoors, d, due to heat gains is difficult to establish because there is no information relating to the presence or otherwise of machinery in the factory. However, there will be lighting and at 500 lux a figure of 15 W/m² will be used assuming industrial trough fluorescent tubes.

$$\text{Temperature rise} \quad d = Q_g/(Q/d_t)$$

$$= (15 \times 2000)/(240\,000/(17+3))$$

$$= 2.5\,K$$

$$\text{Base temperature} \quad t_b = 17 - 2.5 = 14.5\,°C$$

From Table 1.4 DD/SDD = 0.88 by interpolation.

Applying these corrections to the 8-month annual SDD total for Birmingham:

Corrected DD $= 1998 \times 0.875 \times 1.4 \times 0.55 \times 0.88 = 1185$

From Equation 2.2 AED $=$ design load $\times$ (24 DD/d_t)

$$= 240 \times (24 \times 1185/20) = 341\,280\,kWh$$

From Equation 2.3 AEC $=$ AED/seasonal efficiency

From Table 1.2 seasonal efficiency for direct gas-fired heaters is 75%.

$$\text{AEC} = 341\,280/0.75 = 455\,040\,kWh$$

The performance indicator for the factory PI $= 455\,040/2000 = 228\,kWh/m²$

This is well in excess of the benchmark of 107 kWh/m² (Appendix 4) and warrants an energy audit. See Chapter 8 and below.

From Equation 2.4 AFC $= 3.6$ AEC/CV

Calorific value from Table 2.1, CV $= 38.7\,MJ/m³$

$$\text{AFC} = (3.6 \times 455\,040)/38.7 = 42\,329\,m³ \text{ of natural gas}$$

Further consideration to Case study 3.5

The PI for this factory was calculated as $228\,kWh/m^2$ which is excessive when compared with the benchmark of $107\,kWh/m^2$. It is known that the roof which is flat has a thermal transmittance of $2.4\,W/m^2\,K$. By applying $50\,mm$ of insulation lining having a thermal conductivity of $0.035\,W/mK$ to the underside, this can be reduced to $0.54\,W/m^2\,K$. The original heat loss through the flat roof is

$$Q = U \times A \times d_t = 2.4 \times 2000 \times (17 + 3) = 96\,000\,W$$

The revised heat loss through the roof

$$Q = 0.54 \times 2000 \times (17 + 3) = 21\,600\,W$$

The revised design heat loss for the building

$$Q = 240 - 96 + 21.6 = 165.6\,kW$$

The revised annual energy demand is

$$AED = 165.6 \times 24 \times 1185/20 = 235\,483\,kWh$$

$$AEC = 235\,483/0.75 = 313\,978\,kWh$$

Summary of and conclusion for Case study 3.5

- The Performance Indicator for the factory $PI = 313\,978/2000 = 157\,kWh/m^2$
- It should be noted, however, that to avoid the potential risk of interstitial condensation the insulation lining must be impervious to vapour flow.
- This is improved from $288\,kWh/m^2$ but is still above the benchmark of $107\,kWh/m^2$.
- Further consideration would be to ascertain how well the building is sealed since large roller shutter type doors are notorious for increasing the air change rate in a factory.

3.7 The estimation of annual fuel cost for a house

For the reader who has little experience of estimating AEC the final case study considers a domestic residence that hopefully will provide the vehicle for looking at the annual consumption of energy with which the reader is more familiar.

Section 3.9 includes proposals for the next generation of domestic heating systems.

Case study 3.6

A three-bedroom semi-detached family residence on two storeys having a floor area of $100\,m^2$ has a heat loss of $8\,kW$ and is located in Cardiff. Indoor design temperature averaged through the house is taken as $19\,°C$ when outdoor temperature is $-1\,°C$. Estimate the annual energy consumption for space heating and hot water supply from natural gas.

The house is of traditional build with cavity external walls and a ventilated wood on joist ground floor and has three occupants.

Solution

For the purposes of calculation and the given general construction characterisics of the house it is assumed that the building is of medium weight construction. The temperature rise d due to indoor heat gains is limited to $2\,K$ to ensure that a comfortable thermal environment is attained at the expense of energy cost.

Heating is provided initially for 16 hours a day 7 days a week during the heating season.

From Table 3.1 the correction for 7 days occupation per week is 1.0

From Table 3.2 the correction for a length of day of 16 hours is 1.2

From Table 3.3 the correction for a responsive plant and intermittent heating is 0.7

From Table 1.4 the correction for a temperature rise due to indoor heat gains of $2\,K$ which gives a Base temperature of $t_b = 19 - 2 = 17\,°C$ will be 1.18.

Applying these corrections to the annual SDD for Cardiff taking an 8-month heating season from Appendix 1 of 1907, the corrected annual DD for the locality will be:

Corrected SDD $= 1907 \times 1.0 \times 1.2 \times 0.7 \times 1.18 = 1890$ DD

Adopting Equation 2.2 to obtain the annual estimate for space heating the house:

$$AED = \text{design load} \times (24DD/d_t)$$

$$= 8 \times 24 \times 1890/(19 + 1)$$

$$= 18\,144\,kWh$$

The hot water supply for the house must now be considered.

If the house is occupied for 7 days a week, 50 weeks a year, then for a family of three the annual energy estimate for hot water supply can be determined from Equation 4.1 in Chapter 4.

$$AED = \{0.231 \times YN(1 + z/100)EF\}/3.6\,kWh$$

From Table 4.1 the daily use of hot water Y for a medium rental dwelling is 30 litres/person and taking the heat losses as 10% we have for 3 occupants:

$$AED = \{0.231 \times 30 \times 3(1.1) \times 7 \times 50\}/3.6$$

$$= 2223 \, kWh$$

Thus the annual total estimate is $18\,144 + 2223 = 20\,367 \, kWh$
The boiler is fired by natural gas at the rate of 2 p/kWh
Taking seasonal efficiency for a conventional boiler of 75%

$$AEC = AED/(\text{seasonal efficiency})$$

$$= 20\,367/0.75 = 27\,156 \, kWh$$

$$Afc = AEC \times p/kWh$$

$$= 27\,156 \times 2/100 = £544$$

This is equivalent to £45.3 per month

The provision of hot water supply represents $2223/(18\,144 + 2223) = 11\%$ of the annual energy estimate. This accounts for £60 of the annual fuel estimate and is therefore £5 per month.

If the external cavity walls are filled with insulation granules and the house is double glazed resulting in a design heat loss of 6 kW the annual energy cost estimate for space heating and hot water supply will be £423. This is a reduction of 22% in the annual estimate. You should now confirm these figures.

The summary of Case study 3.6 is given in Table 3.4 which includes further analysis.

Table 3.4 Summarising the solution to Case study 3.6

Item	House		Further analysis
	Before insulation	After insulation	
Heat loss	8 kW	6 kW	6 kW
Annual heating	24 192 kWh	18 144 kWh	16 009 kWh
Annual HWS	2964 kWh	2964 kWh	2615 kWh
Annual total	27 156 kWh	21 108 kWh	18 624 kWh
% HWS	11%	14%	14%
PI kWh/m²	272	211	186
Benchmark kWh/m²	247	247	247
Annual heating cost	£484	£363	£320
Annual HWS cost	£60	£60	£52.3
Annual total cost	£544	£423	£372.3
Monthly heating cost	£40.3	£30.25	£26.7
Monthly HWS cost	£5	£5	£4.36
Monthly total cost	£45.3	£35.25	£31
Reduction in annual cost	–	22%	31.5%

Summary of Case study 3.6

• You should now confirm the before and after insulation results in Table 3.4.
• It is worth considering another analysis in which the conventional boiler is replaced with a condensing boiler whose seasonal efficiency is taken as 85%. In order to achieve this it will be necessary to ensure that the heating system has weather compensation control so that the boiler can take advantage of low return water temperatures. The analysis is given in the last column of Table 3.4.
• The HWS consumes an increasing proportion of the annual energy requirement.
• It is likely that if the heating appliances are individually temperature controlled, solar heat gain in the winter season when the sun is low in altitiude and therefore penetrates well into the house will aid in reducing energy consumption.
• Remember that Degree Days do not account for the effects of solar radiation. It is also assumed that the whole house is heated during boiler operating periods to 19 °C when it is likely that the bedrooms will only be heated during part of the operating period and the rest of the dwelling heated to a higher temperature.
• The Building Research Energy Conservation Support Unit (BRECSU) has reviewed some modern ultra low-energy homes in which the total energy delivered is between 113 and 153 kWh/m². You should compare this with the performance indicators in Table 3.4. The benchmark given in the table is taken from Appendix 4.

Refer to Section 11.9 for legislation relating to domestic buildings. Refer to Section 9.11 for an energy audit on a bungalow.

3.8 The estimation of annual fuel cost for a tempered air system

Example 7.3 estimates the annual savings when a recirculation duct is added to an air handling unit operating on full fresh air.

Example 7.4 estimates the running cost for an air handling unit on full fresh air without and with a plate heat exchanger for comparison.

3.9 Reducing fossil fuel consumption in the existing building stock

There is much attention invested in new buildings to ensure that they conform to current Building Regulations and other relevant criteria related to thermal performance and compliance with current benchmarks (see Chapter 11). Refurbishment of existing buildings is also subject to strict compliance. Change of ownership requires that the building may be subject to thermal upgrading. The Climate Change Levy encourages owners of existing commercial and industrial buildings to consider thermal improvement of the building envelope and the upgrade of energy consuming plant. This will be a mandatory requirement to ensure compliance with building benchmarks (see Appendix 4).

3.9.1 Consideration of the existing housing stock

The size of the existing housing stock far outways the annual build of new housing. This sector accounts for around 40% of the UK's CO_2 emissions and presents a serious challenge if the UK is to cut its emissions by 60% by the year 2050.

One proposal for existing low and high rise apartments and dwellings is to convert to community heating by the installation of centralised combined heat and power (CHP) that increases energy efficiency and reduces CO_2 emissions. The electricity generated can be used on site for communal lighting and lifts or directed to the Grid.

For existing detached, semi-detached and terraced housing two proposals are gaining credence for individual systems – both using current technology:

- Solar thermal heating and hot water service employing roof-mounted evacuated-tube collectors, a thermal store, plate heat exchanger for hot water and a condensing combi boiler. A roof mounted plug in domestic turbine is also a serious contender. Alternatively installing six suitably located solar photovoltaic arrays on a three bed semi-detached residence can reduce the household's carbon emissions by more than 20% annually.
- Microchip CHP using the Stirling engine, thermal store and plate heat exchanger. Using thermal storage of water for heating and indirect hot water allows the engine to work at maximum efficiency by running for long periods. Under this regime one unit of energy input will produce 1.55 units of energy output, 45% of which is heat and 55% is electricity.

The electricity generated could be used by the householder or directed into the Grid. This two way communication for electricity would effectively subsidise the cost of heating the homes.

Theoretically at peak times a network of 13 million homes each with this system installed would generate 15 000–23 000 MW of electricity compared with the 15 000 MW produced by the country's ageing nuclear power stations.

Major players in the energy market are looking very carefully at the options including cost, one of which is a leasing arrangement with home owners for centralised CHP and microchip CHP.

Upgrading of this magnitude may have to be considered on change in ownership but it will need substantial financial support if, as the Government White Paper of March 2003 predicted, carbon dioxide emissions must be cut by 60% by the year 2050.

Have another look at Case study 3.6 and suggest further recommendations that, with the incentive of a grant, would substantially reduce reliance on natural gas.

3.10   Further qualifying remarks relating to the case studies

Each of the solutions to the case studies in this chapter must be supported with qualifying remarks. It would be quite unprofessional to submit the fuel estimate calculated in the solution to Case study 3.5 for example without any supporting statements to the client or to senior management. One would be leaving oneself open to criticism and the charge of unprofessionalism. The following is a selection for Case study 3.5

- Auxiliary power for the pumps, fans, controls, fuel burners, etc. is not accounted for in the estimate.
- There may be some machinery in the factory that will require electrical power and influence the heat gains.
- More accurate information relating to the luminaires and level of illumination would be useful.
- No account has been made for hot water consumption.

Can you add to these qualifying remarks?

3.11   Limitations related to Standard Degree Days

More general observations have already been made about the use of SDD. Here again are some reminders:

- The period over which the Annual Degree Days were recorded to obtain SDD for the locality should be stated since they vary from one period to the next and from year to year.
- The place where the building is located may not compare climatically with the region from which the SDD are calculated by the Met. Office.
- Degree Days do not account for the effects of solar heat gain which can be significant through glazing due to the low altitude of the sun in the winter months.
- The chill factor resulting from wind speed is not accounted for in the determination of DD.

3.12 Chapter closure

You now have the skills to estimate the annual energy/energy costs and performance indicators for various types of intermittently heated buildings in the UK and elsewhere in temperate climates if Standard Degree Days are available. The extension of these acquired skills to other building types is a matter of applying the principles learnt in this chapter. It requires reference to current data, access to historical data specific to the building type, if available, and the application of practical common sense. Remember the importance of the Qualifying Remarks relating to the annual fuel estimate and performance indicators in your report.

You are introduced to the potential future for plant and systems for existing housing and the carbon dioxide emissions index or benchmark. Chapter 6 introduces building performance indicators and benchmarks.

Chapter 4

Estimating the annual cost for the provision of hot water supply

Nomenclature

P	probability of usage
N	number of occupants
Y	hot water consumption (litres/person)
E	number of days/week
F	number of weeks/year
AFC	annual fuel consumption (litres, kg, m^3)
AED	annual energy demand (kWh)
AEC	annual energy consumption (kWh)
AFc	annual fuel cost (£)
z	allowance for heat losses (%)
HWS	hot water supply
q	HWS W/m^2 of floor
A	floor area (m^2)
S	number of days in the period under review

4.1   Introduction

The provision of hot water supply for use by the building occupants traditionally has been made either by the adoption of hot water heaters local to the point of use or by central generation via primary hot water connections from the space heating plant to a heat exchanger in the hot water supply cylinder or calorifier.

Current technology now favours separation of space heating from hot water generation and a return to the generation of hot water supply directly in a plant designed specifically for the purpose. This has the effect of increasing both the combustion efficiency and the thermal efficiency of hot water generation and hence a reduction in fuel consumption and emissions and the consequent lowering of operating costs.

4.2   Factors to be considered

The following factors need to be addressed when consideration is given to the provision of hot water supply since they will impinge upon the cost in use.

1   The number and type of fittings, for example personal ablution, laundering, cooking, dishwashing.
2   The number of consumers served.
3   Simultaneous rates of flow that in all probability will require the adoption of the usage ratio P. There are very few applications for the provision of hot and cold water where all the draw-off points on a system of hot water supply will be in use simultaneously. One exception to the rule is groups of showers in a school, clubhouse or factory. The determination of simultaneous rates of flow will influence the heat losses from the pipework and hence the cost in use.
4   Whether fittings are closely grouped or widely distributed. Clearly it is not economic to have long runs of distribution pipework from a central generating source serving fittings widely distributed around the building.
5   Nature of the water supply. This will have a bearing upon the possibility of scale build up and/or corrosion, the need for water treatment and the requirement for maintenance to maintain maximum efficiency of primary energy conversion.
6   Method of generation. If the hot water is generated and stored in a vessel there will be heat losses from the vessel as well as the circulating pipework, which will contribute to the cost in use. The instantaneous generation of hot water on the other hand does not incur this cost and this contributes to the current preference for the provision of direct hot water generation. Heat losses in the circulating pipework will need consideration. Account of the generating method adopted is taken in the choice of seasonal efficiency.
7   Storage/operating temperature. The current consensus is 65 °C as a practical minimum to inhibit the growth of legionella spores. Excessive temperatures increase the effects of scale formation and corrosion. Storage/operating temperature clearly is directly related to costs in use.

Factors 1, 2, 3 and 5 identified above also apply to the provision of cold water supply.

4.3   Hot water supply requirements and boiler power

The requirements for hot water vary in accordance with the type of occupation. Table 4.1 is based upon the plant sizing curves in the CIBSE Guide Book G or the Concise Handbook for a two-hour recovery period.

Table 4.1 Daily hot water consumption and boiler power

Building	Daily water use at 65 °C (litres/person)	Boiler power to 65 °C (kW/person)
Schools		
Service	1.15	0.04
Catering	2.3	0.07
Hotels		
Service	15	0.50
Catering	4.6	0.15
Restaurants		
Service	0.52	0.02
Catering	0.85	0.03
Offices		
Service	1.2	0.04
Catering	2.4	0.08
Shops		
Service	1.9	0.06
Catering	1.6	0.05
Student Hostels		
Service	7.0	0.23
Sports Centres		
Service	20	0.50

The data in the Table 4.1 is for guidance purposes in the absence of historical evidence.

4.4   The annual estimate calculation process

If the daily demand for hot water is Y litres/person for N the number of occupants in the building having access to hot water, then the daily demand will be YN.

If the water is heated from 10 to 65 °C the net daily heat requirement will be:

$$YN \times 4.2 \times (65 - 10) = 231\,YN\,\text{kJ/day}$$
$$= 0.231\,\text{MJ/day}$$

If heat losses are $z\%$ the gross heat requirement will be:

$$0.231\,YN(1 + z/100)\,\text{MJ/day}$$

If the building is occupied for E days per week and F weeks per year, annual energy demand,

$$\text{AED} = \{0.231\,YN(1 + z/100)EF\}/3.6\,\text{kWh} \qquad (4.1)$$

Annual energy consumption,

$$AEC = AED/(\text{seasonal efficiency})\,kWh \qquad (4.2)$$

Annual fuel consumption,

$$AFC = 3.6\,AEC/CV \text{ litres, kg or } m^3 \qquad (4.3)$$

The units to Equation 4.3 depend upon the units for the calorific value CV and therefore will be in litres, kilograms or cubic metres.

Note:

- The conversion is $1\,kWh = 3.6\,MJ$ and therefore $1\,MJ = (1/3.6)\,kWh$
- z, the allowance for heat losses in the circulating pipework and plant is taken as 10% in the following examples. Thus allowance for heat losses:

  $$1 + 10/100 = 1.1$$

- You will see that Equations 4.2 and 4.3 are similar to Equations 2.3 and 2.4.
- Annual fuel cost, AFc, for coal and oil can be calculated from Equation 2.5

  $$AFc = AEC \times \text{cost per litre, kg or } m^3$$

- The annual cost for natural gas can be determined from Equation 2.6

  $$AFc = AEC \times cost/kWh$$

Unlike space heating, the provision of hot water supply will be required throughout the year during occupation periods. It is therefore considered as a "base load." The subject of base loads is considered further in Figure 7.3 and Chapter 10.

4.5 Estimating annual energy consumption

There now follows three examples relating to a school, a factory and a sports centre.

Example 4.1

A new boarding school is to house 210 boarding pupils, 65 day pupils and 20 members of staff. Determine the annual estimate for cost in use related to the provision of hot water supply. Assume indirect heating and a seasonal efficiency of 65%.

Solution

Clearly this is very limited information on which to base an estimate. The length of the terms are needed, for example, and the extent of the sports, cooking, dishwashing and laundering facilities. An allowance will need to be made for occupation outside the term time and the extent to which some of the school's facilities are rented out to other organisations.

Not all the hot water will be generated from the same source; it is possible, for example, that some hand wash basins will be served by local electric water heaters.

If it is assumed that out of the 52 weeks of the year the school is in use for 32 weeks and the data in Table 4.1 is adopted, the following calculations can be made from Equation 4.1:

For the boarding pupils the consumption of 15 litres per day is used from data for hotels in Table 4.1

$$AED = \{0.231 \times 15 \times 210(1.1)7 \times 32\}/3.6 = 49\,803\,kWh$$

For day pupils assuming they attend for 5.5 days a week and using data from Table 4.1

$$AED = \{0.231 \times (1.15 + 2.3) \times 65(1.1)5.5 \times 32\}/3.6 = 2786\,kWh$$

For staff, assuming attendance is averaged out at 6 days per week for 38 weeks per year, Equation 4.1 gives

$$AED = \{0.231 \times (1.15 + 2.3) \times 20(1.1)6 \times 38\}/3.6 = 1110\,kWh$$

Total annual energy demand

$$AED = 49\,803 + 2786 + 1110 = 53\,699\,kWh$$

Annual energy consumption

$$AEC = 53\,699/0.65 = 82\,614\,kWh$$

If the natural-gas tariff is 2.0 p/kWh, annual fuel cost

$$AFc = AEC \times cost/kWh$$
$$= 82\,614 \times 2/100 = \pounds1652$$

Qualifying remarks
When submitting an estimate of this kind it is important to include any qualifying remarks that help to provide the background to the estimate.

- It is necessary to negotiate the tariff with the fuel supplier. It depends upon the quantity of fuel consumed annually, the location of the project and the supplier. Fuel tariffs are considered in Chapter 9.
- This annual fuel estimate is based upon very few hard facts. It therefore needs qualification. You would be open to major criticism if the factors taken in preparing the estimate were not accounted for. Have a look at the qualifications listed in the solution to Example 4.3 and then make a list of qualifications suited to this solution.

The following example considers the hot water requirements for a factory.

Example 4.2
A factory operating on a two-shift system has a total of 120 occupants of which 15 are office staff, 70 are machine operatives and the remainder are manual staff. The office staff do a single nine-hour shift for 5 days a week and half a day on Saturdays. There are two groups of employees who attend the machines and do the manual work, each group undertaking a 10-hour shift 7 days a week. All employees get the statutory holidays and 2-weeks-a-year paid leave.

Estimate the annual fuel consumption from centralised direct oil-fired heaters operating on light grade fuel oil. Seasonal efficiency is estimated at 75% from Table 1.2.

Solution
For the purposes of this estimate the statutory holidays will be discounted. There are three grades of work here namely: office, machine and manual. It will be assumed that the manual workers will require a shower at the end of their shift.

Office staff
The annual number of working days EF for the office staff will be 50 weeks $\times$ 5.5 days and the hot water consumption from Table 4.1 will be used.
Now from Equation 4.1 AED $= \{0.231YN(1 + z/100)EF\}/3.6\,\text{kWh}$
Thus AED $= \{0.231 \times (1.2 + 2.4) \times 15(1.1) \times 5.5 \times 50\}/3.6 = 1048\,\text{kWh}$

Machine operators
The annual number of working days EF per shift will be 50 weeks $\times$ 7 days and for the machine operators an estimate of 10 litres per person per shift will be used in the absence of data.

Thus from Equation 4.1

$$AED = \{0.231 \times 10 \times 70(1.1) \times 7 \times 50\}/3.6$$

$$= 17\,293\,kWh \text{ per shift}$$

For two shifts of machine operators, AED = 34 586 kWh.

Manual workers
For manual workers an estimate of 25 litres per person per shift will be adopted.

Thus from Equation 4.1

$$AED = \{0.231 \times 25 \times 35(1.1) \times 7 \times 50\}/3.6$$

$$= 21\,616\,kWh \text{ per shift}$$

For two shifts of manual workers AED = 43 232 kWh
The total annual energy consumption AED = 1048 + 34 586 + 43 232 = 78 866 kWh
From Equation 4.2 AEC = AED/0.75 = 78 866/0.75 = 105 155 kWh
From Table 2.1 the calorific value for light fuel oil CV = 40.5 MJ/litre
Adopting Equation 4.3 AFC = 3.6(AEC)/CV = (3.6 × 105 155)/40.5 = 9347 litres
Estimated AFC = 9347 litres of light fuel oil.

This is a base load since it is required throughout the year and is not dependant upon the heating season.

Qualifying remarks are again necessary here not least for the estimated water consumption allowed for the machine operators and manual operatives.

The next example considers the hot water requirements for a sports centre.

Example 4.3
A sports centre has the following facilities:

Swimming pool of internationally accepted size, toddlers pool, five squash courts, two basket ball courts, weight training gym, aerobic gym, four games rooms, cafe and restaurant. The centre has five male and female shower rooms to support the sporting activities. It also has five male and female toilets.

Estimate the annual consumption of natural gas associated with the provision of hot water for the centre.

Solution
There will be a number of approaches to this solution. Clearly fuel consumption estimates based upon historical data from a similar sports centre would

prove invaluable here. In the absence of this kind of data one approach is to estimate the occupancy in the various activities offered at the centre and apply a utilisation factor.

Consumption rates for hot water are taken from Table 4.1 where possible. Other rates of consumption are estimated. The data is given in Table 4.2 below to arrive at a daily total hot water consumption for the centre. The column headed "daily use" is an estimate of the number of times or sessions in the day that the facility is used.

The reader should now check the total consumption figures in Table 4.2. For example, for the main pool:

$$\text{Total consumption} = (100 \times 6 \times 20)\, \text{litres/person} = 12\,000\, \text{litres}$$

From the totals the maximum daily occupancy for the sports centre is estimated at 2020. This figure does not include spectators. The total consumption of hot water resulting from this occupancy is estimated as 26 980 litres.

It is unlikely that all activities will be at full capacity on any one day and a utilisation factor of 60% will be adopted here. This reduces the daily provision of hot water to 16 188 litres.

The average consumption of hot water per person will be

$$26\,980/2020 = 13.36\, \text{litres/day}$$

Table 4.2 Data for the solution to Example 4.3

Activity areas	Occupancy estimate	Daily use	Hot water consumption (litres per person)	Total consumption
Main pool	100	6	20	12 000
Toddlers pool	50	6	5	1 500
Five squash courts	10	8	20	1 600
Two basket ball courts	20	8	20	3 200
Weight training gym	20	6	20	2 400
Aerobic gym	30	8	20	4 800
Four games rooms	40	6	5	1 200
Cafe	20	8	1	160
Restuarant	20	4	1	80
Staff	20	2	1	40
Totals: persons per session	330	2020/day	Daily consumption	26 980 litres

For the purposes of this fuel estimate, it will be assumed that the sports centre is open for 51 weeks in the year, 7 days per week and the seasonal efficiency of indirect gas-fired plant is 65%.

From Equation 4.1

$$AED = \{0.231 \times 16188 \times (1.1) \times 7 \times 51\}/3.6 = 407\,910\,kWh$$

From Equation 4.2

$$AFC = 407\,910/0.65 = 627\,554\,kWh$$

From Equation 4.3 and Table 2.1

$$AFC = (3.6 \times 627\,554)/38.7 = 58\,377\,m^3\,of\ natural\ gas$$

It is apparent that there are a number of issues in the solution that could be challenged. In the absence of information specifically relating to the estimate, one must make value judgements based of course on common sense and engineering experience. This is one of the roles of a professional engineer.

The presentation of the fuel estimate to the client or senior management must include the qualifications that have been used in order to justify the estimate. The qualifying statements for this solution are made below.

Summarising solutions to Example 4.3 assuming 60% utilisation
Daily hot water consumption $16\,188/(0.6 \times 2020) = 13.36$ litres/person, AFC 627 554 kWh, annual consumption of natural gas 58 377 m^3, daily consumption of hot water 16 188 litres, estimates based upon utilisation of the sports centre of 1212 persons/day (maximum occupation 2020 persons/day).

Qualifying remarks
- Maximum occupancy at the centre per session is 330 excluding spectators.
- No provision for hot water has been made for spectators.
- A centre utilisation factor of 60% has been applied to estimate the use of the facilities.
- Some of the figures for the consumption of hot water for occupants using the facilities have been taken from recommended guidelines in CIBSE Guide book G. In the absence of data other figures have been based on historical evidence.
- The average consumption of hot water per person at the centre, base is 13.36 litres.
- This annual fuel estimate provides one of the Base Loads for the centre.

- The fuel estimate is for the provision of hot water. It does not include the fuel required to heat make-up water resulting from evaporation from the swimming and toddlers pools. Refer to Chapter 7.

4.6 An alternative method of estimating annual energy consumption

Energy consumption arising from hot water supply can be estimated from knowing the floor area of the building. The CIBSE Building Energy Code tabulates the mean power requirements of hot water supply in W/m^2 for various buildings and this is reproduced in Table 4.3.

Table 4.3 is based upon the provision of hot water supply via a boiler. Annual energy demand for hot water derived from Table 4.3 will be:

$$AED = 24(q \cdot A \cdot S) \, kWh \qquad (4.4)$$

where A = floor area in m^2, S = number of working days under review.

The units of the terms in the formula in which provision of hot water (q) is converted from W/m^2 to kW/m^2 are:

$$h/day \times (kWh/m^2) \times m^2 \times days = kWh$$

There now follows an example relating to a hotel adopting Equation 4.4.

Example 4.4
A hotel has a floor area of $1100 \, m^2$. Estimate the annual energy consumption for hot water supply given that it is open for 11 months of the year. Take seasonal efficiency as 75%.

Table 4.3 HWS provision in W/m^2 of floor

Building type	$HWS/W/m^2$
Office 5- or 6-day week	2.0
Shop 6-day week	1.0
Factories	
5-day single shift	9.0
6-day single shift	11.0
7-day multiple shift	12.0
Warehouses	1.0
Residential	17.5
Hotels	8.0
Hospitals	29.0
Education	2.0

Solution
From Table 4.2 demand for hot water for a hotel is given as $8\,W/m^2$ $(0.008\,kW/m^2)$. If the hotel is in use for 48 weeks in the year, then by adopting Equation 4.4 for annual energy demand:

$$AED = 24 \times 0.008 \times 1100 \times (48 \times 7) = 70\,963\,kWh$$

$$AEC = AED/(seasonal\ efficiency) = 70\,963/0.75 = 94\,617\,kWh$$

The performance indicator for hot water, when the hotel is full, will be:

$$94\,617/1100 = 86\,kWh/m^2$$

You will see from CIBSE Guide book F that the fossil fuel benchmarks for hot water range between 70 and $50\,kWh/m^2$ for types 1, 2 and 3 hotels (Good Practice).

Comparison based on hot water consumption and number of people
It would be helpful to have a comparison with the alternative method of estimating the annual energy demand for hot water. However, to do this it is necessary to know the number of guests that the hotel can house. Adopting an arbitrary figure of $20\,m^2$ of floor per bedroom, which allows for reception, dining and lounge areas as well as the bedroom and access ways in the hotel in Example 4.4, the number of bedrooms at the hotel when it is full will be $1100/20 = 55$.

If each bedroom can take two people there will be a maximum of $2 \times 55 = 110$ guests Adopting Equation 4.1

$$AED = \{0.231\,YN(1 + z/100)EF\}/3.6\,kWh$$

Taking consumption per person from Table 4.1 as $(15 + 4.6)$ litres:

$$AED = \{0.231 \times 19.6 \times 110 \times (1.1) \times 7 \times 48\}/3.6$$

$$= 51\,132\,kWh$$

$$AEC = AED/(seasonal\ efficiency)$$

$$= 51\,132/0.75 = 68\,176\,kWh$$

With a floor area of $1100\,m^2$ the performance indicator for hot water is $68\,176/1100 = 62\,kWh/m^2$ when the hotel is full. This is well inside the benchmarks of $70–50\,kWh/m^2$.

The discrepancy with Example 4.4 is substantial and highlights the importance of using more accurate data or historical data from similar buildings and building use when it is available.

4.7 Chapter closure

You now have the skills to estimate the annual energy/energy costs for the provision of hot water in four categories of building. The extension of these acquired skills to other building types is a matter of applying the principles learnt in this chapter. It requires reference to current data, access to historical data specific to the building type, if available, and the application of appropriate engineering common sense. Remember the importance of qualifying the annual fuel estimate.

Chapter 5

Energy consumption for cooling loads

Nomenclature

SDH   Standard Degree Hours (K·h)
AED   annual energy demand (kWh)
AEC   annual energy consumption (kWh)
d_t design temperature difference (K)
t_{ai} indoor air temperature (°C)
t_{ao} outdoor air temperature (°C)
t_{eo} sol air temperature (°C)
COP   coefficient of performance
t_b Base temperature (°C)
d temperature rise due to indoor heat gains (K)
Q total design cooling load (kW)
Q_g indoor heat gains (kW)
Q_s outdoor solar and conductive heat gains (kW)
EH   equivalent hours of operation at full load (h)
PI   performance indicator (cooling) (kWh/m²)
CFI   cooling flux indicator (W/m²)
CDI   carbon dioxide emission indicator (kg CO_2/m²)

5.1   Introduction

It is not the purpose of this chapter to show how cooling loads are determined. Recourse can be made to another publication in the series for detailed design procedures.

There are four main factors to be accounted for in the design of the maximum cooling load for an occupied building. They are:

- Solar heat gain through glazing, which is instantaneous.
- Conductive heat gain through the opaque envelope of the building. This will be cumulative over time in a building having a medium or high thermal inertia.

- Infiltration of outdoor air.
- Indoor heat gains such as sensible and latent heat gains from the occupants, and heat gains from equipment and lighting. Unlike estimating heat gains for space heating, the occupants are included in the assessment of indoor heat gains for air-conditioning.

To obtain a reasonably reliable estimate of the cumulative effect of conductive heat gain to a given building, it is necessary to use computer-based dynamic simulation. The cooling load for solar and conductive heat gains is calculated from peak conditions in the summer when the refrigeration plant may be required.

Low altitude solar heat gains through glazing in the winter months can be offset by the use of blinds or shading or introducing fresh air. Refrigeration plant should not be used. On this basis the cooling load for external heat gains that is solar and conductive heat gains in the UK therefore can be considered seasonal.

The cooling load required to offset internal heat gains may be required throughout the year if Base temperature is lower than 16 °C.

In most cases, however, Base temperature will be above 16 °C and cooling will not be required for the months of November, December, January and February.

In the winter also the temperature rise indoors after occupation of the building and resulting from internal heat gains will offset the heating load and reduce the demand for space heating.

The total cooling load for a building therefore is the sum of the seasonal or external heat gains Q_s and the indoor heat gains Q_g:

$$Q = Q_s + Q_g$$

The determination of annual energy costs in the UK related to air-conditioning plant providing sensible and latent cooling to buildings is not well documented. The basis upon which estimates of the seasonal cooling load are made for external heat gains is from the use of cooling Degree Hours that can be purchased from the Meteorological Office for the same geographical locations as those for heating Degree Days. Appendix 1 contains Standard Degree Hours for London Heathrow.

There are however limitations on the application of cooling Degree Hours when they are associated with solar and conductive heat gains:

- The air-conditioning cycle can include reheating the air after dehumidification and this is not accounted for.
- The extent of free cooling, like, for example, using ambient air ventilation at night to cool the building structure is not accounted for.

5.2   External heat gains

5.2.1   Cooling Degree Hour data

Appendix 1 includes cooling Degree Hours for London Heathrow for the 20-year period up to 1995 for 18 regions in the UK.

Current emphasis on energy conservation encourages plant operators to consider running the refrigeration plant only during the summer months unless there is a specific requirement for maintaining a constant relative humidity and dry bulb temperature.

5.2.2   Annual energy demand and consumption to offset external and internal heat gains

The annual energy demand for a given design cooling load to offset the effects of solar and conductive heat gain can be calculated from:

$$AED = Q \times SDH/d_t \, kWh \tag{5.1}$$

where d_t = design outdoor temperature (t_{ao}) − design indoor temperature (t_{ai}), $d_t = (t_{ao} - t_{ai})$, $Q = (Q_s + Q_g) \, kW$ and $EH = (SDH/d_t)$ = equivalent hours at full load. See Section 2.3.

Annual energy consumption for electrically operated vapour compression refrigeration central cooling plant:

$$AEC = 0.5(AED) \, kWh \tag{5.2}$$

Annual energy consumption for distributed cooling plant:

$$AEC = 0.78(AED) \, kWh \tag{5.3}$$

The constants for converting AED to AEC are taken from the BSRIA Rules of Thumb and account for the coefficient of performance (COP) of the refrigeration plant and the seasonal efficiency of the electrical plant.

The example which follows is based upon the cooling SDH for London, Appendix 1.

Example 5.1
Determine the annual energy consumption for a distributed cooling plant design load of 150 kW which includes 50 kW of indoor heat gains to a building located in the Thames Valley given that the design temperatures are 30 °C outdoors and 23 °C indoors.

The temperature rise due to indoor heat gains is 5 K.

Solution
The annual cooling Degree Hours for London from Appendix 4 is 3681 hours, for a Base temperature of $t_b = (23 - 5) = 18\,°C$
From Equation 5.1

$$AED = Q \times DH/dt\,kWh = 150 \times 3681/(30 - 23) = 78\,879\,kWh$$

from Equation 5.2

$$AEC = 0.78(AED)\,kWh = 0.78 \times 78\,879 = 61\,526\,kWh$$

Qualifying remark on Example 5.1
- The solution does not account for free night time cooling that may be part of the design; the months of March, April, May and October are included in the annual Degree Hours. If free cooling is available for these months the annual Degree Hours is reduced to 3339 h.

5.3 Factors affecting the estimation of external heat gains

As indicated in the Section 5.1 there are a number of factors in addition to maximum outdoor air temperature, which will affect the annual estimate of energy required to offset solar and conductive heat gains in the summer.

5.3.1 Sol-air temperature

Sol-air temperature, t_{eo}, is defined as the outdoor air temperature which in the absence of solar radiation would give the same temperature distribution and rate of heat transfer through an external wall or roof as exists with the actual outdoor air temperature and the incident solar radiation. It therefore accounts for the high temperatures achieved on the external surfaces of buildings exposed to solar radiation. By definition it also accounts for building orientation since listings are given for vertical and horizontal surfaces.

The effects inside the building on mean radiant temperature are delayed and depend upon the construction of the building envelope. For buildings with massive walls, for example 0.5 m or more of stone, with the outer surface painted a brilliant white and with small areas of recessed glazing, the thermal inertia is considerable and the thermal response factor is high, and the effects of a day of high intensity solar radiation will not be felt indoors until the evening. Conversely a steel frame building with concrete floors and lightweight external wall panels will have a very low thermal inertia and hence thermal response, which will ensure that the effect on mean radiant temperature indoors from solar radiation incident upon the external walls will be almost immediate.

In its refurbishment programme for domestic buildings Germany has used external cladding which has the effect of delaying the increase in mean radiant temperature indoors in the summer until the early evening. The adoption in the UK of this philosophy may allow the use of electricity at a lower night tariff.

The sol-air temperature can be calculated from:

$$t_{eo} = t_{ao} + (R_{so} \times a \times I_t)\,°C$$

where $R_{so} = 0.05\,\text{m}^2\text{K/W}$ for walls and $0.04\,\text{m}^2\text{K/W}$ for roofs; and $a = 0.3$ for light coloured surfaces, 0.6 for medium coloured surfaces and 0.9 for dark surfaces.

The CIBSE Guide gives values of sol-air temperatures for horizontal and vertical surfaces at latitude 51.7N for each month of the year assuming a clear sky.

5.3.2   Conductive heat gains through the building envelope other than glazing

Like heat input to a building during the winter prior to occupation, conductive heat gains resulting from solar radiation take time to penetrate the building envelope. The length of time is dependent upon the thermal inertia of the building and hence its thermal response. It is also dependent upon the location of the thermal insulation sandwich within the building envelope. It was argued that for winter heating the ideal location for an intermittently occupied building was at the inside surface. See Chapter 2.

For conductive heat gains in the summer it can be argued that the best location for the thermal insulation is at the outside surface. This will reduce the impact of solar heat gains indoors and have a beneficial effect on the energy consumption required for seasonal cooling.

It should come as no surprise therefore to frequently find the insulation sandwich located midway in the external walls of buildings in temperate climates.

Conductive heat gains through the opaque element of the building envelope are accounted for in the determination of the maximum design cooling load to offset solar heat gains. Since they are usually cumulative over time this is best determined from reliable software.

5.3.3   Solar heat gain through glazing

Heat gain resulting from solar radiation through glazed windows is instantaneous. The CIBSE Guide gives cooling loads in W/m^2 of glass due to solar gain through vertical glazing at a variety of latitudes and for eight compass orientations for each month of the year.

The BSRIA Rules of Thumb gives approximate figures of $150\,W/m^2$ of glass for south-facing windows from June to September and $250\,W/m^2$ for east-west-facing windows assuming in both cases that internal blinds are in use. This is accounted for in the determination of the maximum cooling load to offset solar heat gains.

The peak seasonal heat gain occurs when the sum of the instantaneous heat gain through the glazing and the heat gain through the opaque portion of the building envelope is a maximum.

5.3.4 Shading

Shading from the effects of direct solar radiation can take a number of forms:

- Shade provided by the effect of recesses in the external envelope of the building
- Shade provided by static or moveable external blinds
- Transient shading provided by the orientation of the building on one or more of its external walls
- Permanent or transient shading provided by surrounding buildings, screens or vegetation.

Permanent fixed and moveable shading should be accounted for in the calculation of the external cooling load to offset solar heat gains.

5.3.5 Reducing external heat gains

For a building whose external envelope is either massive or well insulated thermally, with windows recessed and/or shaded from direct solar radiation, peak indoor mean radiant temperature will be delayed until the evening. If the building is unoccupied at night, an office block for example, the seasonal cooling load can be reduced by the provision of full fresh air mechanical ventilation at night to cool the building structure. The external envelope of course must be designed to delay the effect of the outdoor heat gains on the mean radiant temperature indoors until the evening. If this is done the refrigeration plant will normally only be required to offset the instantaneous solar heat gains through the glazed windows. With suitable external blinds and choice of glass, the peak seasonal cooling load for the building can be reduced substantially. You can see that in this scenario the building structure requires designing with energy conservation in mind. This is not the case with many buildings constructed before 1985 and so the energy manager may well be responsible for an air-conditioned building having a substantial seasonal cooling load which is attributable to solar and conductive heat gains.

The effect of the accumulation of conductive heat gains over time through the opaque envelope of a building will mean that the mean temperature of the structure at the beginning of each day may slowly rise over time for an intermittently occupied building.

5.4   Estimation of the indoor heat gains (Q_g)

Indoor heat gains are calculated from the heat gains within the building. These include sensible and latent heat gains from the occupants, heat gains from lighting and equipment, which is in fairly continuous operation. The BSRIA Rules of Thumb give an approximate guide, based on unit floor area. Table 5.1 lists some of this data:

The heat gains from occupants is better estimated from the number normally in the building and upon the activity, since it can vary from a total of 100 to 400 W. The CIBSE Guide book A gives more detailed data.

There has been a substantial development in energy efficient fluorescent lighting recently using slimline tubes of 16 mm diameter. Claims of 25% less energy consumption at 500 lux illuminance are made and this will clearly have a significant effect upon heat gain per unit floor area attributed to lighting and hence the cooling load.

The energy/facilities manager should consider checking the illuminance levels in the buildings for which he/she is responsible to ensure that they are not excessive for the activities and tasks being undertaken. Appendix 7 tabulates standard service illuminance for various activities and tasks.

Current office equipment such as computers, printers, etc. tend to consume less power and hence contribute less to indoor heat gains than models even 5 years old.

5.4.1   Determination of Base temperature

Base temperature for cooling is that temperature outdoors at and below which the operation of cooling plant is not required.

Table 5.1 Approximate data for the calculation of indoor heat gains

Internal heat gain	Load/unit floor area (W/m^2)
Sensible and latent heat gains from the occupants	20
Lighting	10–25
Office equipment	20–40
Small power	5

Source: BSRIA Rules of Thumb.

With cooling Degree Hours, the lower the Base temperature (indicating higher heat gains) the higher will be the annual Degree Hours and hence the higher the energy consumption.

This is of course different to the Base temperature for space heating where the lower the Base temperature (indicating higher heat gains) the lower will be the heating Degree Days and hence the lower the energy consumption.

Base temperature

$$t_b = t_i - d\,°C$$

where d = temperature rise due to indoor heat gains (K)

and

$$d = Q_g/(Q_s/dt) \tag{5.4}$$

where Q_s = external (solar) heat gains (kW), Q_g = indoor heat gains (kW) and $dt = t_{ao} - t_c$.

5.5 Annual energy consumption – A more detailed analysis

Case study 5.1 takes account of the determination of the temperature rise due to indoor heat gains and finally compares the performance indicator for air-conditioning a Type 3 office with the published Good Practice bench mark.

The performance indicator for the building

$$PI = AEC/(\text{floor area})\,kWh/m^2 \tag{5.5}$$

The cooling flux indicator for the building

$$CFI = Q/(\text{floor area})\,W/m^2 \tag{5.6}$$

Case study 5.1

A refurbished five-storey office block situated in London has a floor area of $450\,m^2$ per floor and has an occupancy of one person per $8\,m^2$ of floor. The heat gains are 130 W/person, lighting $12\,W/m^2$ and those from the office desktop computers which are evenly spread throughout the building are $20\,W/m^2$. The air-conditioning plant is centralised with air cooled condensers positioned on the roof.

Given that the external heat gains to offset peak solar and conductive gains to the building are 80 kW and summer design indoor and outdoor air temperatures are 23 and 28 °C respectively, estimate the annual energy consumption and the performance indicator for the building.

Solution

Number of occupants $= (450 \times 5)/8 = 281$

Heat gain from occupants $= 281 \times 130 = 36.53\,\text{kW}$

Heat gains from lighting and office equipment in the case study are taken from Table 5.3

For permanent artificial lighting, heat gain $= 12 \times 450 \times 5 = 27\,\text{kW}$

Heat gains from office equipment $= 20 \times 450 \times 5 = 45\,\text{kW}$

Total heat gains indoors $= 36.53 + 27 + 45 = 108.53\,\text{kW}$

From equation 5.4, $d = 108.53/(80/(28 - 23)) = 7\,\text{K}$

Base temperature $t_\text{b} = 23 - 7 = 16\,°\text{C}$

From Appendix 4 the annual cooling Degree Hours for London must be interpolated from Base temperatures of 18 and 12 °C. Assuming a linear association this is calculated as $8235\,\text{K} \cdot \text{h}$.

You should now confirm agreement with the annual degree hours.

The total cooling load $Q = Q_\text{s} + Q_\text{g} = 80 + 108.53 = 188.53\,\text{kW}$

From Equation 5.1

$$\text{AED} = 188.53 \times 8235/(28 - 23) = 310\,509\,\text{kWh}$$

From Equation 5.2 for centralised plant

$$\text{AEC} = 0.5 \times 310\,509 = 155\,255\,\text{kWh}$$

From Equation 5.5

$$\text{PI} = 155\,255/(450 \times 5) = 69\,\text{kWh/m}^2$$

Conclusions relating to Case study 5.1

*   The annual energy cost of cooling from electrically operated plant at 10 p/kWh will be

$$\text{AEc} = 155\,255 \times 10/100 = £15\,526$$

*   For a Type 3 office the good practice benchmark from Appendix 4 is $52\,\text{kWh/m}^2$. This compares with $69\,\text{kWh/m}^2$ for the offices. As it is a refurbished building the offices should be at or below the good practice benchmark. It would be worth considering here one of the methods of free cooling to reduce the building performance indicator.
*   The cooling flux indicator for the building from Equation 5.6

$$\text{CFI} = 188\,530/(450 \times 5) = 84\,\text{W/m}^2$$

- The carbon dioxide indicator using the factor for electricity from Table 6.1 will be

$$\text{CDI} = \text{PI} \times 0.43 = 69 \times 0.43 = 30 \, \text{kg} \, CO_2/\text{m}^2$$

Annual carbon dioxide emissions $= 30 \times (450 \times 5) = 67.5 \, \text{tonnes}$

- The benchmark emissions for carbon dioxide $= (52 \times 0.43) \times (450 \times 5) = 50.3 \, \text{tonnes}$.

Qualifying remarks for solution to Case study 5.1
- In the winter the space heating plant will be required to bring the building up to temperature before occupation.
- The external cooling load, Q_s, of 80 kW and the internal cooling load, Q_g, of 108.53 kW will be required during the summer months.
- Indoor heat gains Q_g in the winter will offset some of the heating load.
- The effects of cooling load Q_g to offset indoor heat gains during the winter season could be predicted from historical data from the energy consumption for space heating.
- It is left to you to list further qualifying remarks relating to estimating annual energy consumption for vapour compression refrigeration cooling plant in occupied buildings. For example, what information would be needed to generate a thermal performance line for refrigeration plant responding to indoor and outdoor heat gains. The independent variable will be Degree Hours and the dependent variable will be Energy Consumption. Have a look at Chapter 10.

5.6 Free cooling

The free cooling chiller is not new. It specifically refers to package air cooled chillers that have an integral free cooling circuit with a self contained control system. In addition to the compressor, evaporator and condenser there is an additional dry cooler circuit. The chilled water is cooled or partially cooled with ambient air resulting in a subsequent reduction in the requirement for direct expansion cooling from the refrigerant.

The condenser fans also serve the dry cooler circuits and using intelligent control they control the compressor head pressure whilst maximising the free cooling when it is available.

5.6.1 Cost savings in use

The Enhanced Capital Allowance scheme (ECA – refer to Chapter 11) for packaged air cooled chillers sets a qualifying energy efficiency ratio (EER) or coefficient of performance (COP) of 2.6 for a 500 kW chiller. Since

COP = (cooling duty output)/(electrical input), the electrical input to a chiller of this size will be $500/2.6 = 192\,\text{kW}$.

If the chiller is installed in a data centre or a high use IT office where the load is constant throughout the year the approximate annual running cost at 7 p/kWh will be:

$$192 \times 365 \times 24 \times (7/100) = £117\,734$$

Checking the units of terms:

$$\text{kW} \times (\text{days/year}) \times (\text{h/day}) \times (\text{cost/kWh}) = \text{cost/year}.$$

The use of the free cooling chiller will typically reduce the annual cost by about 35%.

5.6.2   Designing a free cooling system

The free cooling system in the chiller must be operated with a water–glycol mix to ensure against the risk of freezing.

Clearly the higher the design chilled water temperatures the higher the energy savings. A chiller operating at water flow and return temperatures of 10 and 16 °C will gain a higher savings benefit than operating temperatures of 6 and 12 °C.

5.7   Chapter closure

You have been introduced to cooling Degree Hours issued for one locality and how they are calculated. You know the limitations of applying this data to the estimation of cooling loads for occupied buildings. The determination of the temperature rise due to indoor heat gains and the calculation of Base temperature have been considered. Building performance indicators have been determined and compared with published benchmarks.

Chapter 6

Performance Indicators

Nomenclature

PI	Performance indicator (annual) (kWh/m^2, $kg\ CO_2/m^2$)
AEC	annual energy consumption (kWh)
AED	annual energy demand (kWh)
Benchmarks (annual)	(kWh/m^2, kWh/bedroom)
EUI	Energy use indices (annual) (kWh/m^2)
CEI	Carbon dioxide emission indices (annual) (kg/m^2)

6.1 Introduction

Local authorities and other owners of large numbers of widely dispersed buildings and even owners of buildings on single sites have the opportunity of determining performance indicators for the purpose of comparing the performance of buildings with similar characteristics.

These provide one of the checks and balances which help to promote authenticity and validation in the preparation of accounts for building stock to auditors and shareholders.

If for example the maintenance costs on a particular building type prove to be excessive when compared with another building being used for the same purposes and located in a similar environment, or when compared to appropriate historical data, it provides evidence for future policy action. If no comparison is made no one is any the wiser and additional costs continue to accumulate.

Energy consumption in buildings is considered in the same manner and buildings having similar usage and occupancy patterns can be compared. Historical data is now available in the form of tables of benchmarks which express annual energy consumption per square metre of floor for good practice and typical practice and for different building types.

The Climate Change Levy and Government legislation have imposed tough new requirements on the part of building owners who now have to show that an energy management policy is in place and energy performance indicators for the buildings for which they are responsible are compared with the published benchmarks and improved upon by setting targets.

GOOD PRACTICE and TYPICAL PRACTICE benchmarks are published in the CIBSE Guide book F and Action Energy's Energy Consumption Guides. Refer to Appendix 4.

6.2 Performance Indicators

For existing buildings there are two ways of determining the performance indicator for space heating, both of which should be adopted in an analysis.

- Adopting SDD for the building location and any appropriate corrections to arrive at a performance indicator.
- Determining the performance indicator for the building from recent historical fuel invoice records.

Clearly the latter will give the factual indicator whereas the former will offer an estimate. However, the former may give an insight into the prevailing indicator. These two indicators can then be compared with the appropriate benchmark. There is likely to be sufficient evidence from these data for further analysis of the building performance.

For the purposes of this book the term "performance indicator" is that determined for the building being analysed by the building owner or representative and the term "benchmark" is that obtained from published data for comparison.

The performance indicator for a building is calculated from:

$$PI = AEC/\text{treated floor area kWh/m}^2$$

Performance Indicators are also determined from gross floor area. Reference to the published data in CIBSE Guide book F identifies which is to be used. Treated floor area excludes plant rooms and other areas not heated. Annual energy consumption (AEC) refers to the historical data from fuel invoices.

For a projected building fuel invoices are not available and annual energy demand (AED) can be determined from design data (see Chapters 2, 3 and 4) and the estimated AEC = AED/seasonal efficiency. The units for measurement are kWh. Note that $1\,kWh = 3.6\,MJ$. This conversion is required when a fuel calorific value is used, commonly in MJ/m^3, kg or litre.

You should now look at the remaining case studies in Chapter 3 in which this calculation has been done.

Clearly, calculated performance indicators must be compared with care with the published benchmarks since they will depend upon many factors.

- From Chapters 1, 2 and 3 it is apparent that performance indicators for projected buildings will be affected by the local climate or microclimate that may vary from the region in which the Met Office SDD are taken.
- They will vary in accordance with occupancy patterns given in Tables 3.1 and 3.2.
- SDD will vary with the classification of building and plant, see Table 3.3.
- They may also be adjusted in response to indoor heat gains. Refer to Example 1.1.

Likewise for an existing building where historical fuel accounts can be used to arrive at a performance indicator there will be factors to consider when comparing the performance indicator with the benchmarks.

The determination of the performance indicator for an existing building may well be part of the Energy Audit (see Chapter 8) and requires information about:

- Fuel accounts for the previous 2 or 3 years
- Architects drawings of the building
- Occupancy patterns
- Knowledge of the existing space heating and domestic hot water systems, lighting and power use
- Whether or not the occupants are currently satisfied with the indoor temperature conditions
- Standards of maintenance of the space heating systems and auxiliary services.

Some of this information can be extraordinarily difficult to obtain particularly if the building is years old. However, if there is difficulty in locating the fossil fuel accounts and floor plans in particular an audit will take longer to complete.

The second performance indicator for a building is that related to carbon dioxide emissions where:

$$PI = (PI)\,kWh/m^2 \times emission\ factor$$

This gives building performance in $kg\,CO_2/m^2$ floor.

Emission factors for various fossil fuels are given in Table 6.1.

Government and other authorities use the term "tonnes or kilograms of carbon". This is easily converted to tonnes or kilograms of carbon dioxide as follows: For example

$$kg\,C/m^2 = (kg\,CO_2/m^2) \times (12/44)$$

Table 6.1 Emission factors for fossil fuels

Energy source	Emission factor kg CO_2/kWh
Electricity (grid)	0.43
Coal	0.29
Natural gas	0.19
Fuel oil	0.27

The ratio of (12/44) comes from the chemical equation by mass for burning carbon to carbon dioxide: $C + O_2 \rightarrow CO_2$

$$12 + 32 \rightarrow 44$$

6.3 Building benchmarks

Historical data relating to energy use in various buildings as provided in the CIBSE Guide book F provides useful benchmarks for comparisons with building performance indicators calculated by the building owner or facilities manager, and, more importantly, can be used to show improvements or otherwise in the energy performance of the building under review.

6.3.1 Classification of levels of acceptance

The published benchmarks are listed as GOOD PRACTICE and TYPICAL. The aim for most existing buildings is to have lower consumption figures than the "good practice". However, following an energy audit inherently inefficient buildings may not be improved to this standard for practical and economic reasons. In such cases a realistic target that lies between "good practice" and "typical" benchmarks should be adopted.

The "good practice" benchmarks should be used as upper limits for new buildings that should have calculated performance indicators well below the benchmark.

Having calculated the performance indicator and compared it with the published benchmark, it is important to set a sensible target PI for the building at the next period of review.

6.4 Further analysis of Case study 3.5

The reader should now refer to the factory in Case study 3.5 that clearly has a high PI. Following this case study is an analysis of a roof upgrade and its effect on the PI. Clearly the cost of lining the roof of the factory must be taken into account. For example given a projected cost of the lining as £10/m² the cost estimate would be $10 \times 2000 = £20\,000$.

The saving in AED $= (U_1 - U_2) \times A \times 24 \times DD \, \text{kWh}$

Checking the units of terms, AED $= \text{kW/m}^2\,\text{K} \times \text{m}^2 \times \text{hours/day} \times \text{K} \times$ days $= \text{kWh}$

Note that the thermal transmittances must be in $\text{kW/m}^2\,\text{K}$

$$\text{AED} = (0.0024 - 0.00054) \times 2000 \times 24 \times 1185 = 105\,797 \, \text{kWh}$$

$$\text{AEC} = \text{AED/seasonal efficiency}$$

$$\text{AEC} = 105\,797/0.75 = 141\,062 \, \text{kWh}$$

Taking the cost of natural gas as 2 p/kWh, cost saving $= £2821$ per annum.

Simple payback $= (\text{cost of improvement})/(\text{net saving}) = £20\,000/2821 = 7$ years.

A decision now has to be made on whether or not to go ahead. The advantages are:

- The annual cost saving in fuel after 7 years during which fuel costs will rise.
- A reduction in the building performance indicator from 288 to $157 \, \text{kWh/m}^2$.
- An improvement in comfort for the factory occupants and
- There may be a consequent increase in production.

6.5 Carbon dioxide emissions

Conversion factors from energy in kWh to kg of carbon dioxide emission are based upon that for year 2000+ emission factors.

Clearly there is a direct correlation between energy consumption and carbon dioxide emission; a saving in the annual use of fossil fuel will produce a corresponding reduction in carbon dioxide emissions. Table 6.1 gives emission factors for various fuels.

Note that the factor for electricity depends on the mix of fossil fuels and renewable energy in its generation. The figure given is the recommended current value.

6.6 Annual carbon dioxide emission benchmarks/indices (CDI)

From Table 6.1 and Appendix 4 carbon dioxide emission indices for space heating can be calculated and are given in Table 6.2 for four building types.

Table 6.2 Carbon dioxide emission benchmarks/indices for space heating

Building type	Carbon dioxide performance in $kg\,CO_2/m^2$ Good practice		
	Coal	Oil	N/Gas
Schools; secondary (no pool)	31	29	20
Offices; Types 1&2	23	21	15
Industrial pre-1995: up to $5000\,m^2$	31	29	20
Residential	72	67	47

6.7 Further analysis of Case studies 3.1 and 3.2

Case study 3.1
The four-storey office has a calculated PI of $146\,kWh/m^2$ and an HFI of $80\,W/m^2$. From Appendix 4 the good practice benchmark is $79\,kWh/m^2$ and if the office is new build the PI should be well below this benchmark. The carbon dioxide emission performance indicator can be calculated for the office from Table 6.1 where for fuel oil the emission factor is 0.27. Therefore

$$PI = 146 \times 0.27 = 39.42\,kg\,CO_2/m^2$$

This compares with the good practice benchmark of $21\,kg\,CO_2/m^2$ from Table 6.2.

Clearly it follows that a poor performance indicator for energy is followed by a poor performance indicator for carbon dioxide emission. You will notice that the heat flux index for the building is $80\,W/m^2$ and therefore the design heat load for the building of $115\,kW$ needs to be reduced.

As this is a new project it is not too late to improve the thermal performance of the building envelope.

Case study 3.2
The six-storey office is to be refurbished and like new buildings it is subject to Building Regulations Approved Document L.

The proposed space heating system gives an energy performance indicator of $77\,kWh/m^2$ and a heat flux indicator of $58.6\,W/m^2$.

From Table 6.1 the carbon dioxide emission factor for natural gas is 0.19.

The emissions performance indicator for the refurbished building will be:

$$PI = 77 \times 0.19 = 14.63\,kg\,CO_2/m^2$$

These indices compare well with the published good practice benchmarks of $79\,kWh/m^2$ from Appendix 4 and $15\,kg\,CO_2/m^2$ from Table 6.2.

6.7.1   Performance indicators for low energy buildings

The monthly Journal of the Chartered Institution of Building Services Engineering has recently included in its building analysis articles on new and refurbished buildings, annual energy targets in kWh/m^2 of treated floor. This provides a useful update which accounts for recent changes in the standards of thermal insulation of the building envelope. Table 6.3 gives details of the energy targets from some of the buildings analysed.

The Building Research Energy Conservation Support Unit (BRECSU) has reviewed the performance specifications for a new office building and seminar facilities at the Building Research Establishment's site at Garston in which performance targets are identified. These are shown in Table 6.4.

Table 6.3 Performance Indicators in kWh/m^2 for buildings analysed

Building	Heating	Ventilating	Hot water	Small power	Lighting	Total
Charities Aid Foundation, Kings Hill, Kent *Refrigeration*	100	30	18	25	45	218
New Scottish Office, Edinburgh	107	9	3	14♦	26	159
RSPB office, Bedfordshire *Ventilation*	–	–	–	–	–	140
Marston Book Services, Oxfordshire	12•	24	•	1.2	1.72	39
Learning Centre, East Anglia University	–	–	–	–	–	95
Elizabeth Fry building, East Anglia University	26•	•	•	8	16	50
No. 1 Leeds city office park	45•	16	•	48♦	11	120

Notes
1 Symbol • identifies services whose energy targets have been combined in the building analysis.
2 Symbol ♦ identifies small power as including lifts.
3 The European Commission's grant scheme EC2000 funds eight non-domestic buildings two of which have been completed in the UK. The scheme requires a 50% reduction in energy and carbon dioxide emission compared with traditional buildings and no air conditioning.

Table 6.4 Performance indicators for a new office and seminar facility, BRE

Building type	Annual energy consumption (kWh/m^2)	Annual carbon dioxide emission (kg/m^2)
Narrow plan	Gas 47, elecricity 36	34
Narrow plan	All electric 68	46
Deep plan	Gas 47, electricity 43	39
Deep plan	All electric 75	51

In both Tables 6.3 and 6.4, the low annual energy targets (below $70\,kWh/m^2$) generally indicate buildings which have been specifically designed as low energy buildings taking advantage of solar heat gains and using natural ventilation for air replacement and night time cooling.

6.8   Chapter closure

You are now able to determine performance indicators from source material for different building types using data from the fossil fuel accounts and comparing with the Performance Indicators determined from SDD.

Following the determination of the building's energy and carbon dioxide performance indicators and taking care in interpreting the results you can advise a client on the projected performance for a new or refurbished building or the current performance in the case of an existing building and show how it compares against published benchmarks.

You will also need to advise on what can be done to bring the energy performance of the building in line with the current Building Regulations Approved Document L for new and refurbished buildings.

You understand the importance of taking into account any local factors when determining the performance indicator as they will influence a comparison with the published benchmarks.

Chapter 7

Energy conservation strategies

Nomenclature

CEM	contract energy management
PEM	partnership energy management
EH	Equivalent hours at full boiler load
n_c	heat conversion efficiency
n_u	heat utilisation efficiency
n_s	seasonal efficiency
AEC	annual energy consumption (kWh)
AED	annual energy demand (kWh)
PI	performance indicator $(kWh/m^2, kg\,CO_2/m^2)$
db	dry bulb temperature (°C)
M	mass flow rate (kg/s)
d_h	difference in enthalpy in kJ/kg dry air
v_{fr}	volume flow rate (m^3/s)
v	specific volume (m^3/kg)
d_g	difference in moisture content (kg/kg dry air)
E	exhaust air
O	outdoor air-condition
S	supply air-condition
R_s	room stat
D_s	duct stat
OD	outdoor detector
ID	immersion detector
M	meter, kg/s and million
C	controller
A	actuator
ρ	density (kg/m^3)
C	specific heat capacity (kJ/kg K)

7.1 Introduction

Since the present national average for the purchase of fossil fuel for heating is 4% of turnover, the incentive on the part of senior management to invest in energy saving measures is low with many organisations. The demands upon time are considerable and, when cost cutting is required, other areas of investigation that offer potentially greater savings than, say, 10% of 4% of turnover take precedence.

However, pressure is now being applied throughout much of industry via the Climate Change Levy and on the domestic scene to reduce the levels of greenhouse gas. Pressure for the reduction of carbon dioxide and oxides of nitrogen may well encourage building owners into action to reduce energy consumption.

The Audit Commission's long-established guidelines on energy management state that for every £1 million of fuel purchased one energy manager should be appointed and 10% of the capital expenditure on fuel reinvested in energy conservation measures. This hardly encourages those organisations that have smaller fuel bills. Many such organisations have no energy management policy and those that have give the job to an employee who has other responsibilities. In such circumstances the individual given the part-time job of energy manager needs all the encouragement that he or she can find from top to bottom of the organisation. The large organisations typically have the management of energy consumption as company policy. It is clearly in the interests of the British Airports Authority, for example, with $186\,000\,m^2$ of floor space to have strategies for energy management. Some large organisations like local authorities and hospitals buy in the energy management expertise in the form of partnership energy management (PEM) or contract energy management (CEM), in which the contractor takes on responsibility for operating and updating the services and plant on the site or sites and shares the savings in energy consumption with the client. With PEM the contractor offers the same services but allows the client to have day-to-day control on plant operation.

7.2 Energy transfer from point of extraction to point of use

For fossil fuels there are a number of stages from extraction to site use that need consideration. Energy conservation strategies apply to every stage although it is at the site use stage where this book is focused.

Figure 7.1 is a block diagram showing the stages from extraction to site use. The non-productive energy use identifies the energy losses sustained in extraction, refinement and transport.

Figure 7.2 is a block diagram showing the energy use on site. The non-productive loss of energy on site is identified as the energy losses due to

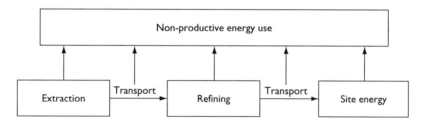

Figure 7.1 Processes in production and distribution of fossil fuel.

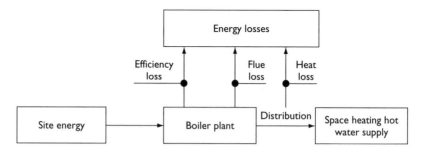

Figure 7.2 Energy losses from point of supply to point of use.

the inefficiency of the boiler plant and that via the sensible and latent heat losses from the flue gases. If the heating medium transports the heat energy via external ducts there will be losses here also.

Primary energy is that in the fossil fuel at the point of extraction. Site energy is that in the fossil fuel available to the consumer on a site. Table 7.1, which originates from data published by CIBSE, lists primary energy fuel factors which relate directly to units of site energy. Note the size of the primary fuel factor for electricity due to the inefficient conversion of fossil fuels in electrical power only generating plant.

Table 7.1 Primary energy fuel factors

Final energy form	Primary energy fuel factor	Remarks
Electricity	3.82	Power only generating plant
Manufactured fuels	1.38	
Oil	1.09	
Natural gas	1.07	
Coal	1.03	

Combined heat and power plant, for example, increases the conversion efficiency from about 30% for power only generating stations to between 50 and 80% for CHP plants. This will have the effect of reducing the fuel factor to between 1.5 and 2.0 Part (f) of the following example shows how the primary energy fuel factor in Table 7.1 is applied.

Example 7.1

The owner of a light engineering factory pays £27 273 for a year's supply of light fuel oil for the heating boiler.

a How many litres of oil is delivered to site per annum at a cost of 30 p/litre?
b What is the annual energy consumption of the fuel in MWh?
c If the boiler consumes 30 litres of oil per hour, what is its power input in kW?
d Find the equivalent hours of operation at full boiler load EH.
e If the average efficiency of the boiler between services is 70%, what is the useful power output in kW?
f How many units of primary energy does the boiler consume for each unit of useful energy output?
g If the floor area of the factory is 3000 m^2 find the performance indicator for the building and compare with the benchmark.

Solution

a The quantity of fuel oil burnt $= £27 273/0.30 = 90 909$ litres.
b AEC of the fuel $= 90 909 \times 11.26 = 1 023 635$ kWh $= 1024$ MWh. Note the energy per litre for light fuel oil of 11.26 kWh is obtained from Table 9.2
c Boiler input $=$ litres of oil/h $\times$ energy per litre $= 30 \times 11.26 = 337.8$ kW.
d Equivalent hours of operation at full load $= 90 909/30 = 3030$ hours/annum
e Useful boiler output $=$ power input $\times$ efficiency $= 337.8 \times 0.7 = 236$ kW.
f From Table 7.1 the primary energy content of each unit of site fuel is 1.09. For each GJ of site energy used 0.7 GJ is useful energy. Thus each useful GJ of energy requires $1/0.7 = 1.43$ GJ of site energy.

    The boiler will therefore consume 1.43 GJ of site energy for each GJ of useful energy.

    The boiler will therefore consume $1.43 \times 1.09 = 1.56$ GJ of primary energy from the point of extraction for each GJ of useful energy.
g The performance indicator for the factory PI $=$ AEC/(floor area) $= 1 023 635/3000 = 341$ kWh/m^2.

    This compares with the benchmark of 286 kWh/m^2, which is equivalent to a shift pattern of work of 1.7 taken from Appendix 4. Further examples of building performance are given in Chapter 6.

7.3 Efficiency of space heating plants

There are three types of efficiency related to space heating systems namely: heat conversion efficiency, utilisation efficiency and seasonal efficiency. The CIBSE Guide book F tabulates these efficiencies for different types of plant and system.

7.3.1 Heat conversion efficiency (n_c)

This relates to the boiler plant where $n_c = $ (output)/(input), where output = heat delivered to the system (kW) and input = heat potential in the fuel (kW).

Heat conversion efficiency should in fact be reasonably steady to a turn down ratio of 30% of full load. At lower loads boiler efficiency falls away rapidly. It should be noted that heat conversion efficiency is that quoted by the boiler manufacturer and is determined under controlled conditions and not by prevailing conditions on site. Heat conversion efficiencies vary for different boilers from 98% for double condensing boilers to 85% for conventional boilers.

7.3.2 Utilisation efficiency (n_u)

This addresses the manner in which heat is emitted where it is needed. It is dependant upon the type of system, the method of temperature control, the disposition and sizing of the space heating appliances, the size, construction and thermal behaviour of the building and the method of operation.

$$n_u = \text{(design heat load)/(heat delivered to the system)}$$

7.3.3 Seasonal efficiency (n_s)

This is the overall efficiency of boiler plant and system over the heating season and accounts for variations in efficiency over time resulting from the intervals between maintenance and variations in the load on the boiler plant. Refer to Table 1.2.

Typical seasonal efficiencies vary from 60% for indirect heating like low temperature hot water to 75% for direct heating like direct fired gas heaters for water or air at the point of use.

The relationship between heat conversion efficiency (n_c), utilisation efficiency (n_u) and seasonal efficiency (n_s) is expressed as:

$$n_s = n_c \times n_u$$

7.4   Seasonal and base load demand and consumption

In order to apply energy conservation strategies it is necessary to distinguish between demand and consumption and between the terms "seasonal" and "base load".

Seasonal demand and consumption varies with changes in climate over the year and is associated with space heating and air conditioning.

Base load demand and consumption relates to a constant and continuous use of energy or fuel as in the case of hot water supply and some manufacturing processes.

The difference between demand and consumption can be illustrated by distinguishing between annual energy demand, AED, and annual energy consumption, AEC:

$$AEC = AED/n_s$$

or rearranging:

$$AED = AEC \times n_s$$

Figure 7.3 shows a typical plot of seasonal and base load energy consumption for the provision of space heating and hot water supply to a building for 12 months. The plot is made from monthly fuel readings taken on site.

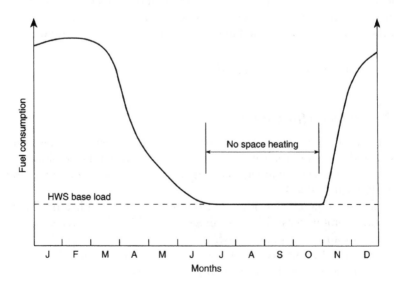

Figure 7.3 Fuel consumption profile for a typical year for space heating and hot water supply.

In practice there may be small variations in the base load each month. The plot gives a picture of the consumption pattern of fuel over a year. If there is only one metering point for the fuel the size of the base load can be identified during the summer months when space heating is not required. The base load line can then be extended either side of the summer months on the reasonably safe assumption that it will be more or less constant.

The plot gives little more information and does not show plant performance.

7.5 The energy conservation programme

Good reasons should be given for senior management to agree upon a programme of energy conservation. In fact there are at least two good motives for action on the part of building owners/occupiers and these will become increasingly effective in time. One is cost saving on reduction in consumption of fossil fuel and the other is to reduce carbon dioxide emissions via the products of combustion into the atmosphere. The Climate Change Levy will provide the motivation for the latter via grants and penalties. Increases in the tax on fossil fuels will encourage the former.

Manufacturing industry and other large users of fossil fuels in the UK are already some way along the path of energy conservation and the reduction in emissions of carbon dioxide. This is due mainly to the effect of the Climate Change Levy and significant achievable savings on fuel costs resulting from deliberate changes in working practices and modernisation of energy consuming plant.

The area which has still to respond and where much work is still needed is in the commercial, local authority and domestic sectors where fossil fuels are used for space heating, cooking and hot water supply only. Here the potential savings on fuel costs have not always been seen as significant as those in industry. Once the nettle has been grasped a standard procedure of checks and balances can be adopted for an energy conservation programme.

A flow diagram best illustrates the strategies in the programme and is shown in Figure 7.4.

7.5.1 Initial commitment

As a result of potential future rising fuel costs and current legislation on emissions of carbon dioxide, the attention of building owners will begin to turn towards the cost of fuel. The commitment will therefore eventually come from senior management in the organisation. This is very important since the energy conservation programme will affect all those working in the organisation and will cost money to implement. Commitment to a programme of energy conservation currently is sluggish in the areas identified and needs the impetus of highly motivated and enlightened individuals.

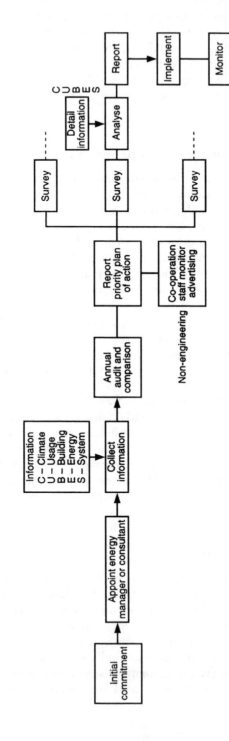

Figure 7.4 General flowchart of energy conservation programme.

7.5.2 Appointment of the energy manager/facilities manager

This may be an internal appointment, or a consultant may be appointed from outside the organisation. Either way since energy conservation is about changing things and managing change, the individual must have authority and be non-partisan. The energy manager's responsibilities would be:

- Initiating the energy conservation programme in the organisation
- Undertaking an energy survey and energy audit on site
- Analysing the energy use on site and reporting to senior management
- Preparation of cost and payback of identified conservation measures
- Implementing the agreed conservation measures
- Monitoring the results and comparing with forecasts
- Agreeing tariffs with the energy suppliers and purchasing the energy.

The responsibility of the energy purchaser for the organisation might be seen as the first and only job that the energy manager need undertake. Judicious purchasing of energy will have the most immediate impact on fuel costs for the organisation. However, purchasing is a one-off phenomenon and does not obviate the need for meeting future energy targets since the cost of fossil fuel will rise substantially over time.

7.5.3 Collecting appropriate information

Having obtained official support from senior management who must advise all the staff to assist in his/her endeavours, the energy manager then needs to collect all relevant information relating to the building's energy and use. This can be done under the following initials which relate to the flow diagram in Figure 7.4:

C   Climate: Degree Days for locality, mean daily outdoor temperatures, hours of day light, hours of sunshine

U   Usage: hours of occupancy, cleaning hours, holiday periods, special use, overtime periods, shifts, population, product output, water consumption

B   Building: general state of repair of the external envelope, site plans, floor plans, elevations, U values, modifications, age, life

E   Energy: fuel consumptions from fuel bills for coal, gas, oil, bottled gas, electricity, maximum demand values, tariffs, standing charges. Two to three years data is required

S   Systems: system layouts, design loads, manufacturers outputs, nameplate data, modifications, controls schematic.

7.6   The energy survey

The energy survey should result in an understanding of the energy flows within the buildings and identify energy wastage. A survey will have four principal objectives:

- To identify the points of energy input, conversion, distribution and use
- To identify the factors which are likely to affect the level of energy use for each area or item of plant
- To identify areas of energy wastage and where efficiency could be improved by changing working practices, better maintenance, upgrading or replacement
- To identify whether there is a programme of preventative maintenance for the building and its services.

Following the survey it should be possible to analyse the information and data from all sources and suggest means for improving efficiency and reducing wastage, and calculate the level of potential savings for each case.

It may be that the instigation of a programme of preventative maintenance is the first task to be achieved by the energy manager. It is all too frequently not done properly.

The three main activities in the energy survey are:

- Locating the energy input and metering points, boiler plant, water heaters, etc.
- Energy conversion efficiency checks, responsibilities, etc.
- Energy utilisation: tracking the converted energy, is it used efficiently, how is it controlled, responsibilities, etc.

Figure 7.5 shows a schematic diagram of a space heating system serving radiators, unit heaters and the heater battery in an air handling unit. Applying the three activities of the energy survey to the schematic diagram of the space heating system:

i   Metering the fuel supply to the boilers, metering the power supply to the pumps, fans and controls.

ii Metering the heat supplied to the three circuits. If this is not possible because of lack of metering equipment, clamp on thermometers can be used to find the flow and return temperatures at the circuit connections on the boiler header and at the points of use.

iii Checking the boiler combustion efficiency test and its frequency.

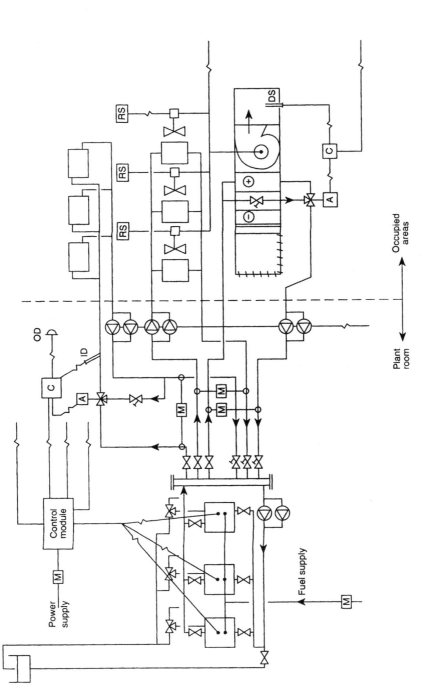

Figure 7.5 Schematic diagram for the energy survey.

iv Checking that the space heating appliances are operable: Are the areas in which the space heating appliances are located occupied when the heating circuit is on.

v Checking the frequency of cleaning to the unit heaters and heater battery coils and the filters on the air handling unit.

vi Checking the temperature controls and settings.

vii Checking the drives on motors to pumps and fans. Checking the frequency of flushing out the heating system.

viii Checking the thermal insulation to the distribution pipework.

You can see from the activities undertaken here that the four principal objectives in the survey can now be addressed.

It is likely, however, that the only activity which comes out as hard data in this survey is the meter-reading on the fuel supply and that obtained from the fuel invoices which would have been read by the fuel supplier. The energy manager may therefore have something to say at this stage in the survey report.

In the absence of heat meters or flow meters, flow and return temperatures to each circuit can be checked using contact thermometers. See Appendix 5.

Some of the checks for the system shown in Figure 7.5 may appear to be basic. However, they are important and some should form part of the preventative maintenance programme for the services within the building anyway. If a building energy management system is installed some of the checks can be undertaken at the central station or outstation but a visit to each part of the building is essential.

7.7   The energy audit

This is discussed in detail in Chapter 9. However, it is important to ascertain the extent and cost of energy use on site. From the floor plans, the floor area can be determined and the performance indicator for the building calculated in kWh/m^2.

An initial comparison with the benchmarks for the building type can be made.

An energy audit allows the following matters to be aired:

• Understand the scale of energy use and its annual cost.
• Compare the energy use with Benchmarks.
• Make initial proposals on potential energy saving measures that will meet the benchmark conditions.
• Cost the proposals.
• Cost the potential savings on the fuel accounts.

7.8   Areas for energy saving

Refer to Appendix 2 for a comprehensive list.

There are three areas which can be considered namely: the building, the services within it and the usage requirements for the building.

7.8.1   The building

Thermal insulation of the external envelope and reduction of unwanted infiltration of outdoor air are the two main factors. The latter will include the use of air locks at entrances, the use of self-closing doors in stairwells and on corridors, well-sealed windows, control of ventilation in lift shafts. The former will be helped by having a recognised programme of external maintenance.

7.8.2   The services

There are two factors here: services provided for the occupiers and not requiring their intervention like space heating, ventilation, air conditioning and services used by the occupier like hot and cold water supply, lighting, etc.

Here are some areas for potential energy saving in the provision and use of services.

- Space heating: efficiency of generation, efficiency of distribution (pump power and thermal insulation), efficiency of utilisation, time control, temperature control.
- Hot water services: temperature of supply, quantity of supply.
- Ventilation: quantity of fresh air, time control, temperature control, efficiency of distribution (fan power).
- Electrical power: control of use, correct tariffs and maximum demand.
- Air conditioning: free cooling, fan control, time, temperature and humidity control.
- Heat recovery: use and application of heat recovery devices such as thermal wheels, heat pipes, run around coils, plate heat exchangers, condensing boilers.
- Energy saving equipment: variable speed pumps and fans, weather compensated temperature controls, direct-fired non-storage hot water supply plant, boiler water temperature controls.

You should make yourself familiar with heat recovery and energy saving equipment available on the market.

7.8.3   Building occupation patterns

If time control for space heating is considered for a moment an office having a nine to five day for five days a week over a thirty-nine week heating season requires heating for 1560 hours. The total time over that period is 6552 hours. The heating is therefore required for only 24% of the total time. The thermal insulation in the external envelope should therefore be close to the inner surface to keep the preheat period to a minimum, Mondays requiring the longest preheat. Plant run times are important for energy conservation.

7.9   Heat recovery

One of the ways in which energy can be conserved is by the provision of heat recovery plant. As with any form of energy conservation the feasibility of introducing heat recovery equipment must be investigated. The following are key parameters:

   i   There is a sufficient quantitiy of recoverable heat.
   ii   The recovered heat can be made available at a suitable temperature.
   iii   There is a use for the recovered heat.
   iv   The waste heat source and point of re-use are not too remote.
   v   There is a match between the time of heat demand and the time of waste heat availablity.

   Some examples of successful heat recovery would include:

- Heat recovery from the flue gases on space heating boiler plant, Example 7.2.
- The use of the recirculation duct on a ventilation/air-conditioning system, Example 7.3.
- The use of a plate heat exchanger in the return and fresh air intake ducts on a full fresh air ventilation/air-conditioning system, Example 7.4.
- Recovery of the sensible and latent heat from the return duct of swimming pool ventilation plant, Examples 7.5 and 7.6.

Example 7.2
The plant servicing a heating system with weather compensated temperature control providing constant volume variable temperature to circuits of radiators is to be replaced.
   The new plant will consist of one condensing boiler and one conventional boiler. The cost of replacement is estimated at £4600. From the data, determine the estimated saving in fuel and the simple payback period.

Data
Annual cost of natural gas £4308; efficiency of original boiler plant 70%
Efficiency of new boiler plant is 90% made up of an efficiency for the new
conventional boiler of 85% and that for the condensing boiler of 95%
Charge for natural gas 2 p/kWh

Solution
Note that the efficiencies quoted are the heat conversion efficiencies provided by the boiler manufacturers. It is not necessary to consider seasonal
efficiencies but it is important to compare like with like.

If each of the new boilers operates for 50% of the time, the average
efficiency can be taken as $(95 + 85)/2 = 90\%$

> The annual energy consumption, AEC $= 4308 \times 100/2 = 215\,400\,\text{kWh}$
> The annual energy demand, AED $= 215\,400 \times 0.7 = 150\,780\,\text{kWh}$
> The new AEC $= 150\,780/0.90 = 167\,533\,\text{kWh}$
> The estimated annual cost for gas $= 167\,533 \times 2/100 = £3351$
> The estimated annual saving $= £4308 - £3351 = £957$
> Simple payback $=$ capital cost of measure/annual saving
> The estimated simple payback $= 4600/957 = 4.8$ years

Cost benefit analysis which includes simple payback is investigated in
Chapter 8.

Example 7.3
A tempered air mechanical ventilation system supplies $4.5\,\text{m}^3/\text{s}$ of air to a
building located in London. Occupancy is 5 days a week, 12 hours a day.
The ventilation plant is taken as responsive and the building is of medium
weight.
It is decided to undertake a feasibility study to consider introducing air
recirculation in the air handling plant which currently operates on full
fresh air.

Data
Occupancy is 160, indoor design temperature is $19\,°C$, outdoor design temperature is $-2\,°C$, minimum fresh air per person is 8 litres/s, air density is
$1.2\,\text{kg/m}^3$, specific heat capacity of air is $1.02\,\text{kJ/kg K}$, seasonal efficiency
of boiler 70%, cost of natural gas is 2.1 p/kWh, capital cost of recirculation
duct, mixing dampers and controls is estimated at £5500.
It is assumed that the pressure developed by the existing extract fan will
cope with the altered ductwork.

Solution
> From Appendix 1 the annual SDD is 2033.
> From Tables 3.1, 3.2 and 3.3 the correction factors appropriate here,
> interpolated where necessary, are 0.8, 1.13 and 0.7 respectively.

Corrected DD $= 2033 \times 0.8 \times 1.13 \times 0.7 = 1287$
Equivalent hours of operation at full load EH $= 24\, \mathrm{DD}/d_t$
For an explanation of this operator refer to Section 2.3
EH $= 24 \times 1287/(19+2) = 1471$
Minimum fresh air supply $= 160 \times 8/1000 = 1.28\,\mathrm{m^3/s}$
Saving in design heat load $\mathrm{d}Q = \mathrm{d}(v_{\mathrm{fr}}) \times \rho \times C \times d_t\,\mathrm{kW}$
Checking units of terms $= (\mathrm{m^3/s}) \times (\mathrm{kg/m^3}) \times (\mathrm{kJ/kgK}) \times \mathrm{K} = \mathrm{kJ/s} = \mathrm{kW}$
Then substituting: $\mathrm{d}Q = (4.5 - 1.28) \times 1.2 \times 1.02 \times (19+2) = 83\,\mathrm{kW}$
Energy saved by return air recirculation in the heating season:
Estimated saving in annual energy demand in $\mathrm{kWh} = \mathrm{kW} \times \mathrm{EH}$

$\mathrm{d(AED)} = 83 \times 1471 = 122\,093\,\mathrm{kWh}$

Estimated saving in annual energy consumption

$\mathrm{d(AEC)} = 122\,093/0.7 = 174\,419\,\mathrm{kWh}$

Estimated saving in annual fuel cost

$\mathrm{AFc} = \mathrm{d(AEC)} \times \mathrm{cost} = 174\,419 \times 2.1/100 = \pounds 3663$

Estimated simple payback $=$ capital cost of measure/annual saving $=$
$\pounds 5500/3663 = 1.5$ years.

Since the return on the investment is so short the scheme is financially feasible. It would then be necessary to confirm the practical feasibility of the proposal.

Example 7.4
An alternative proposal is now considered for the tempered air system serving the building in Example 7.3.

One of the disadvantages of air systems incorporating recirculated air is that air contaminated with bacteria which cannot be cheaply filtered out is returned and inhaled by the occupants unless sophisticated and expensive air filtrarion is used. A continuous filtered fresh air supply is preferable and will assist in the elimination of the potential for sick building syndrome.

The proposal is to consider the feasibility of using a plate heat exchanger as a recuperator between the fresh air supply and the return ducts.

Data
Air pressure drop in both the exhaust and the fresh air sides of the recuperator is 190 Pa.
Heat transfer efficiency of the recuperator is 64%.
Fan efficiency is 68%, fan drive efficiency is 98%, fan motor efficiency is 92%.

Charge for electricity is 10.0 p/kWh.

Capital cost of the recuperator including the extra costs to the supply and
extract fans and motors in overcoming the increased resistance to air
flow, insulation and controls is £10 000.

Solution

The connections to the recuperator are shown in Figure 7.6. Note the bypass
arrangement for use outside the heating season.

Equivalent hours of operation at full load = 1471

Saving in design heat load $dQ = n_s \times v_{fr} \times \rho \times C \times dt \, \text{kW}$
$$= 0.64 \times 4.5 \times 1.2 \times 1.02 \times (19 + 2)$$
$$= 74 \, \text{kW}$$

Estimated annual energy saving = $dQ \times EH = 74 \times 1471 = 108\,854 \, \text{kWh}$

Estimated annual cost saving $d(\text{AFc}) = (108\,854/0.7) \times 2.1/100 = £3266$

Fan power consumed in overcoming the air resistance in the recuperator =
$2 \times 190 \times 4.5/0.68 \times 1000 = 2.515 \, \text{kW}$

Operating hours for the fans = 5 days a week $\times$ 12 hours a day $\times$
39 weeks = 2340 hours

Note: Outside the heating season the recuperator is bypassed.

Electrical energy used = $2.515 \times 2340/0.98 \times 0.92 = 6527 \, \text{kWh}$

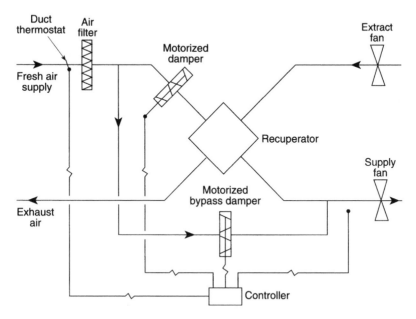

Figure 7.6 Application of a plate heat exchanger recuperator in supply and extract
ducts.

Estimated annual charge $= 6527 \times 10/100 = £653$
Estimated net annual saving $= £3266 - £653 = £2613$
Estimated simple payback $= £10\,000/£2613 = 3.83$ years.

Clearly the payback period is longer than that required for the provision of recirculated air. However, it provides a better quality of ventilation than the use of recirculated air.

Further analysis of the tempered air system in Example 7.4

Using the data from Example 7.3 and the application of the plate heat exchanger the annual fuel consumption before and after its installation can be estimated.

From Equation 2.2 $\text{AED} = Q \times 24 \, DD/d_t = Q \times EH \, kWh$

Before installation

For air flow $Q = v_{fr} \times \rho \times C \times dt \, kW$
Taking typical values for $(\rho \times C) = 1.2 \times 1.02 = 1.224 \, kJ/m^3$,

$$Q = 4.5 \times 1.224 \times (19 + 2) = 115.7 \, kW$$

Checking units of terms $= (m^3/s) \times (kJ/m^3 K) \times K = kJ/s = kW$

$$\text{AED} = Q \times EH = 115.7 \times 1471 = 170\,195 \, kWh$$

From Equation 2.3 $\text{AEC} = 170\,195/0.7 = 243\,135 \, kWh$
The annual fuel cost before installation, from Equation 2.5, is

$$\text{AFc} = 243\,135 \times 2.1/100 = £5106$$

After installation

$$\text{AED} = Q \times EH \times (1 - n_s) = 115.7 \times 1471 \times (1 - 0.64) = 61\,270 \, kWh$$

$$\text{AEC} = 61\,270/0.7 = 87\,529 \, kWh$$

$$\text{AFc} = 87\,529 \times 2.1/100 = £1838$$

Saving in cost $= 5106 - 1838 - 653 = £2615$

Estimated simple payback $= 10\,000/2615 = 3.82$ years

This agrees with the earlier calculation.

In Example 7.4 only the sensible heat in the exhaust air is reclaimed. The enthalpy of a typical sample of room air might comprise 60% sensible heat and 40% latent heat. If you were considering two heat recovery appliances, one operating on sensible heat reclaim and the other on total enthalpy and both had a stated efficiency of 70%, the latter would give a genuine 70%

recovery but the former would recover only the sensible heat from the room air. Thus its efficiency will be 70% of 60% which is 42%.

Example 7.5

A swimming pool hall is supplied with full fresh air at 4.72 m³/s and 35 °C db, 10% saturated. The outdoor air is sensibly heated from saturated conditions at −1 °C. The design conditions for the pool hall are 28 °C db and 70% saturated which are also the air conditions in the extract duct.

Determine the sensible and latent heat content of the exhaust air and the rate at which moisture is being lifted from the pool.

Determine also the design load on the air heater battery.

Solution

Figure 7.7 shows the winter cycle on a sketch of the psychrometric chart.

You will need a copy of the CIBSE psychrometric chart to make full sense of the solution.

The specific enthalpy of the air exhausted to atmosphere 71.5 kJ/kg.
The specific enthalpy of the fresh air entering the air handling unit is 7.5 kJ/kg

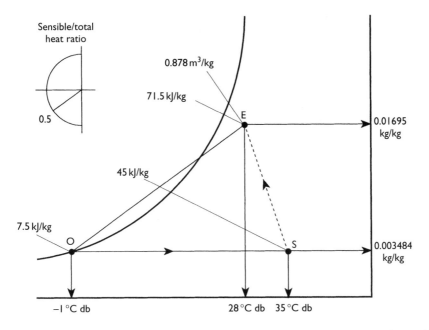

Figure 7.7 Psychrometric cycle without the thermal wheel: supply air at 35 °C db, 10% saturated; exhaust air at 28 °C db, 70% saturated (Example 7.5).

The heat content of the exhaust air $= M \times d_\text{h} = (v_\text{fr}/v) \times d_\text{h}$

$$= (4.72/0.878) \times (71.5 - 7.5)$$

$$= 344\,\text{kW}$$

From Figure 7.7 the sensible to total heat ratio is about 0.5:1.0 and therefore the sensible heat content will be $344/2 = 172\,\text{kW}$ and the latent heat content is $172\,\text{kW}$.

The moisture lifted from the pool will be $= (v_\text{fr}/v) \times d_\text{g}$

$$= (4.72/0.878)$$

$$\times (0.01695 - 0.003484)$$

$$= 0.07239\,\text{kg/s}$$

$$= 261\,\text{kg/hour}.$$

Clearly there is opportunity for heat recovery here, from both the sensible and the latent heat from the exhaust air and the loss of treated and heated water from the pool.

If the pool is in use for 12 hours a day with the air handling plant operating, the water loss will be $261 \times 12 = 3132\,\text{kg}$ per day.

The design output of the air heater battery for the pool hall will be from point O to point S on Figure 7.7:

Air heater battery output $= (v_\text{fr}/v) \times d_\text{h} = (4.72/0.878) \times (45 - 7.5)$

$$= 202\,\text{kW}$$

Substantial savings in energy demand and make-up water for the pool can be achieved with the use of a thermal wheel. This is explored in Example 7.6, which follows.

Example 7.6
Consider the installation of a total enthalpy thermal wheel for the air handling plant serving the pool hall in Example 7.5 and identify the potential savings during a winter cycle. The manufacturers of the thermal wheel claim an efficiency of 70%.

Solution
Figure 7.8 shows a thermal wheel and Figure 7.9 shows the winter cycle on a sketch of the psychrometric chart. Note that the exhaust condition is now the condition of the air extracted from the pool hall before it is exhausted to atmosphere.

It is important to ensure that the moisture in the air does not condense out during the process of total heat recovery. A preheater therefore is required

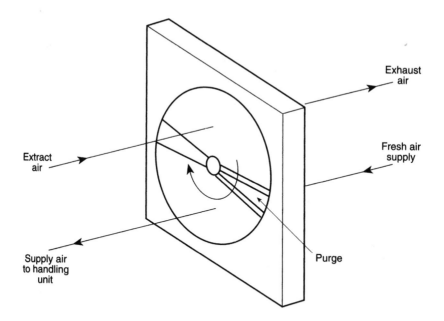

Figure 7.8 The thermal wheel (motor and drive omitted).

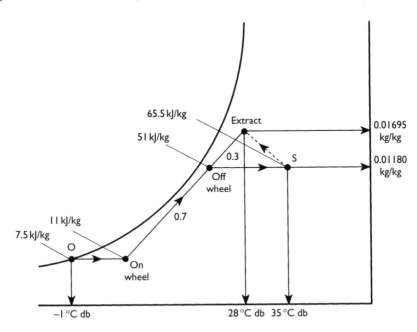

Figure 7.9 Psychrometric cycle with the thermal wheel at 70% efficiency (Example 7.6).

to sensibly raise the temperature of the incoming fresh air from the design saturated condition of $-1\,°C$ db to about $2.5\,°C$ db, on wheel condition. This prevents the line on the chart between outdoor condition O and the extract condition from crossing the saturation curve.

The Off wheel condition will be 7/10 along the line On wheel to the Extract. An after heater is needed here to sensibly heat to supply condition S. The combined outputs of the preheater and after heater will be less than the air heater battery required in Example 7.5 so there is already a saving here.

> The heat content of the exhaust air $= 344\,kW$
> Referring to Figure 7.9, the energy saved by the total enthalpy thermal wheel $= (4.72/0.878) \times (51 - 11) = 215\,kW$
> The heat content of the exhaust air $= 344 - 215 = 129\,kW$
> Heat recovery from the exhaust air $215/344 = 62.5\%$
> The design load on the preheater will be $= (4.72/0.878) \times (11 - 7.5) = 19\,kW$
> The design load on the after heater will be $= (4.72/0.878) \times (65.5 - 51) = 78\,kW$
> The total sensible heating load at design conditions $= 19 + 78 = 97\,kW$

This load compares with the design load of $202\,kW$ in Example 7.5 before the use of the thermal wheel.

This represents a potential saving on sensible heating of $(202 - 97)/202 = 52\%$

The moisture lifted from the pool is also reduced offering a further saving:

$$\text{Moisture lifted} = (4.72/0.878) \times (0.01695 - 0.01180)$$
$$= 0.02769\,kg/s$$
$$= 100\,kg/hour$$

Comparing with Example 7.5 for a 12-hour period the moisture lifted from the pool is now $1200\,kg/day$ instead of $3132\,kg$. This represents a saving in make-up water, water treatment and heating for the pool of 62%.

Summarising Examples 7.5 and 7.6
The design data before and after the proposed use of the total enthalpy thermal wheel is tabulated in Table 7.2.

When you check the solutions to Examples 7.5 and 7.6 you may find small discrepencies in reading off the specific enthalpy values from the psychrometric chart. The savings indicated will be achieved only when the outdoor design condition is saturated air at $-1\,°C$ db. It would be reasonable therefore to assume that these represent maximum savings. A more realistic approach would be to determine the savings from an average outdoor winter condition for the locality.

Table 7.2 Summary of solutions to Examples 7.5 and 7.6

Item	Before	After	Saving (%)
Rate of heat discharge in exhaust	344 kW	129 kW	62.5
Design sensible heat load	202 kW	97 kW	52
Water loss from pool	3132 litres/day	1200 litres/day	62
Make-up pool water treatment	3132 litres/day	1200 litres/day	62
Make-up pool water heating	3132 litres/day	1200 litres/day	62

The use of the total enthalpy thermal wheel begs serious consideration, however.

Only the winter cycle has been analysed here and it would be appropriate to consider the operation of the plant during the summer. You should now consider the summer cycle taking outdoor air at, say, 27 °C db, 70% saturated. Is additional air handling plant required?

If you are in doubt you should seek advice. Alternatively recourse can be made to another book in this series.

Consider now the alterations required for the installation of the thermal wheel. It would be necessary to have the fresh air intake and exhaust ducts located adjacently at a point upstream of the proposed preheater so that the thermal wheel can be positioned astride the two air streams. See Figure 7.8. This and the replacement of the air heater battery and the increased resistance to air flow will require substantial alterations to be made to the air handling plant with bypass ducts around the thermal wheel.

It is now suggested that you consider the potential energy savings using an average outdoor winter condition (for the Thames Valley) of 6.5 °C db 80% saturated.

The solution is tabulated below in Table 7.3 and the cycle is shown on a sketch of the psychrometric chart in Figure 7.10.

Table 7.3 Summary of final solution to Example 7.6 using an outdoor condition of 6.5 °C db 80% saturated.

Item	Before	After	Saving (%)
Rate of heat discharge in exhaust	284 kW	81 kW	71
Design sensible heat load	150 kW	73 kW	51
Water loss from pool	2817 kg/day	871 kg/day	69
Make-up water treatment	2817 kg/day	871 kg/day	69
Make-up water heating	2817 kg/day	871 kg/day	69

Notes
 i For an average outdoor condition of 6.5 °C db no preheating is required. The preheater must be in place, however, to satisfy the requirements at the design outdoor condition of saturated air at −1 °C db.
 ii Contrary to the initial assessment the savings look potentially better than for operation at the design outdoor condition.

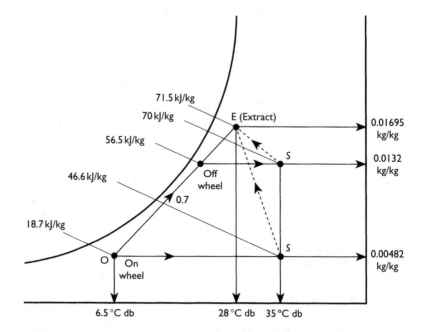

Figure 7.10 Psychrometric cycles with and without the thermal wheel; outdoor condition 6.5 °C db, 80% saturated.

7.10 Chapter closure

You now have the tools to proceed with an energy survey that is underpinned with the strategies, checks and balances accounted for in the chapter. Energy audits are briefly introduced and examples are presented in Chapter 9. Some work has been done on the application of heat recovery equipment. You will have seen therefore how useful it is to be knowledgeable in the design of heating, ventilating and air-conditioning systems. Recourse can be made, relating to system design, to other publications in this series.

Chapter 8

Cost–benefit analysis

Nomenclature

SPB	simple payback in years
TV	terminal value (£)
P	principal sum (£)
r	rate of interest (%)
n	period of term in years
PV	present value (£)
R	return on investment
DF	discount factor
CPV	cumulative present value factor
PVA	present value of an annuity
PWF	present worth factor
PVA = CPV = PWF	
TVF	terminal value factor
NPV	net present value (£)

8.1 Introduction

It is a fact of life that much of the way modern society organises itself requires accountability. This, it might be argued, is a fine sentiment because it provides a check and a balance upon error and excess. In the current market place economy, one of the requirements of accountability, means cost accounting.

Some enlightened building owners have, however, taken a step change in whole life costs by including in the accounting system environmental and sustainable issues promulgated by the Government commission chair Sir Jonathan Porritt.

A business philosophy is emerging which recognizes that not only can care for the environment in its broadest sense be profitable but it can result in a very good marketing strategy.

On completion of this chapter you would be well advised to consider this step change in philosophy in relation to the whole life costs of the services in a new or refurbished building and how environmental and sustainable issues might be factored into the business accounts. Whole life costs are discussed further in Chapter 12.

However, the matters that concern us here are those relating to operating costs and management of fossil fuel consumption in building services. The energy manager, consultant or facilities manager has to show senior management or the client that the costs of saving energy (which will contribute even if it is in a small way to an increase in the quality of the environment) are met by cost savings in the consumption of fossil fuel and hence a reduction in the expenditure column of the company's annual balance sheet. The Climate Change Levy and rising fuel costs will provide the focus here. There are various ways of looking at the benefits of cost accounting measures for energy saving.

In some of the examples so far in the text simple payback is used. This is acceptable if inflation and interest rates are low and the period of payback is short. Some organisations prefer simple payback and simple costing of alternative energy saving schemes because of the vaguaries of future inflation and interest rates. However, the client or investor may want to compare the potential return on an investment in a proposal for energy saving with the potential return on investment in another aspect of the business. Whole life costing also requires a more sophisticated approach.

For this reason other ways of costing the benefits attributable to operating costs and energy saving proposals are considered in this chapter and compared with simple payback.

8.2 Simple payback

The formula is:

$$SPB = (\text{cost of energy saving proposal})/$$
$$(\text{annual saving} - \text{annual cost of saving})$$

The annual cost of the energy saving proposal might for example be the cost of electricity for operating the energy saving equipment and/or the annual cost of maintenance on the energy saving proposal.

Have a look at Case study 2.1 in which simple payback and discounted payback have been calculated.

8.2.1 Length of the payback period

Payback periods of up to 5 years are usually acceptable. It is a current fact of life that longer periods of payback may not be popular, much in the same way as long-term investment does not immediately attract the potential investor.

Clearly the payback period must be less than the life of the energy saving measure and less than the life of the building and the services installation as a whole.

8.3 Discounted cash flow and present value

Consider an investment of £100 at an interest rate (rate of return or discount rate) of 5%.

After the first year the sum would become $100 \times 1.05 = £105$
After the second year the sum would become $100 \times (1.05)^2 = £110.25$

In general, after n years the original sum P invested would become the terminal value and $TV = P(1 + r/100)^n$, where r is the percentage rate of annual interest or rate of return.

Alternatively we can say that the present value PV of £110.25 at the end of year 2 is worth £100 today if discounted at 5%.

Thus

$$PV = P(1 + r/100)^{-n}$$

Substituting we have

$$PV = 110.25(1 + 5/100)^{-2} = £100$$

The following case study looks at discounted cash flow.

Case study 8.1
Consider a situation where £3000 is spent out of an organisation's reserves on an energy saving measure which has no further cost and the annual savings in fuel are £1000. The simple payback period will therefore be $3000/1000 = 3$ years. Undertake a discounted cash flow and present value analysis and compare the results.

Solution
If the money spent on the energy saving measure had instead been invested for 3 years at a discount rate of 5%, the terminal value would be:

$$TV = 3000(1 + 5/100)^3 = £3473$$

and the return R on the investment will be:

$$R = 3647 - 3000 = £473$$

The return on the investment in the energy saving measure at the end of the fourth year will be £1000 net, and £1000 net annually thereafter for the life of the measure as long as fuel prices remain in line with general inflation.

There are a number of points which can be made here:

- There is clearly no return on the investment in the energy saving measure during the first 3 years as the annual savings are paid for the original investment.
- The charge, if there is one, for servicing the loan of £3000 must be accounted for as well.
- If the life of the energy saving provision is 20 years the net savings will accrue for the remaining 17 years and amount to £17 000 at present value.
- If the energy saving measure had not been undertaken and the £3000 left in the organisation's reserves, after 20 years at 5% it would be worth £7960 providing a return of £4960. There is clearly a strong motive to prosecute the energy saving provision.
- The net savings accruing from the energy conservation measure are £17 000 - 4960 = £12 040

An alternative approach is normally adopted in which the payback period accounts for the initial loss in earnings. The results of the energy saving provision are given in Table 8.1.

Table 8.1 Tabulated solution to Case study 8.1

Start of year	Cost of measure (£)	Energy saving (£)	PV factor	PV (£)	Cumulative PVs (£)
0	3000	–	–	−3000	−3000
1	–	1000	0.953	+935	−2065
2	–	1000	0.907	+907	−1158
3	–	1000	0.864	+864	−294
4	–	1000	0.823	+823	+529

Notes
i Present value factor $= 1/(1+r/100)^n = (1+r/100)^{-n}$ (8.1)
ii PV = energy saving × PV factor
iii A table of PV factors is to be found in the Appendix 6 for ascending values of r and n.
iv Some sources employ the discount factor which is $(1+r/100)^n$ in which case the present value is obtained from PV = energy saving/discount factor.

From the results in Table 8.1 we see in the cumulative PV column that although simple payback showed that the initial outlay would be paid for in 3 years, the effect of the interest rate means that it takes between 3 and 4 years before the energy saving measure is paid for.

It is possible to determine accurately what the payback period will be from the following formula:

$$\text{Cumulative present value factor CPV} = (1 - (1 + r/100)^{-n})/(r/100)$$

$$(8.2)$$

The cumulative present value factor is also known as the present worth factor (PWF) and present value of an annuity (PVA). A table of PVA factors is given in Appendix 6 for ascending values of r and n. Also

$$\text{CPV} = \text{simple payback period}$$

Therefore in this case study

$$3 = (1 - (1 + 5/100)^{-n})/(5/100)$$

from which

$$0.15 = 1 - (1.05)^{-n}$$

and

$$(1.05)^{-n} = 0.85$$

Taking the reciprocal on both sides of the equation,

$$(1.05)^{n} = 1.176$$

Taking logs to the base 10:

$$n(\log 1.05) = \log 1.176$$

therefore

$$n = \log 1.176 / \log 1.05$$

and payback period

$$n = 3.331 \text{ years.}$$

Initial summary of Case study 8.1

- Simple payback takes 3 years for the original cost of the energy saving provision to be paid.
- Discounted payback takes 3.33 years for the original investment to be paid. This accounts for the loss in revenue which would otherwise accrue by leaving the £3000 invested at an interest rate of 5%.
- Now have a look at Case study 2.1 in Chapter 2, in which CPV is used.

8.4   Effects of fuel inflation (Case study 8.1)

It is likely that to reduce levels of carbon dioxide emissions, building owners will need an incentive to reduced consumption of fossil fuel in their building stock. This is now embodied in the Climate Change Levy that will eventually affect all building owners. Taxing the use of fossil fuel will also provide an incentive. Taking Case study 8.1 and working on the assumption of nil general inflation but an annual inflation on fossil fuel of 6%, the results as shown in Table 8.2 will occur.

It can be seen from the cumulative present value column in the tabulated results that the measure will pay for itself in about 2.9 years.

Table 8.2 Accounting for fuel inflation in Case study 8.1

Start of year	Cost of plant (£)	Energy saving (£)	Energy inflation	Cash flow (£)	PV factor	PV (£)	Cumulative PV (£)
0	−3000	–	–	−3000	1.0	−3000	−3000
1	–	1000	1.06	+1060	0.952	+1009	−1991
2	–	1000	1.06^2	+1124	0.907	+1019	−972
3	–	1000	1.06^3	+1191	0.864	+1029	+57

Note
1 Cash flow = energy saving × annual energy inflation
2 Annual energy inflation is obtained from the terminal value factor:

$$\text{TVF} = (1 + r/100)^n \tag{8.3}$$

3 A table of terminal value factors is given in Appendix 6 for ascending values of r and n.

Final summary of Case study 8.1

Capital cost of measure (£)	Annual saving (£)	Payback period	Method of calculation
3000	1000	3 years	Simple payback
3000	1000	3.33 years	5% discount
3000	1000	About 2.9 years	5% discount and 6% annual fuel inflation

As one would expect progressive inflation of fossil fuel prices gives a better return on the capital investment in energy saving measures by reducing the payback period and hence provides the incentive on the part of building owner. The difficulty lies in assessing what the future fuel inflation and interest rates will be.

Three formulae have been introduced in this section:

PVF $= (1 + r/100)^{-n}$ – This formula accounts for depreciation or inflation (Equation 8.1).

CPV $= (1 - (1 + r/100)^{-n})/(r/100)$ – This formula enables annual repayments on a loan to be calculated. It also enables the calculation of the present value of annual costs over a fixed term (Equation 8.2).

TVF $= (1 + r/100)^n$ – This formula allows the calculation of compound interest on a single sum invested over a fixed term. It is also used here to calculate the annual energy inflation on fuel (Equation 8.3).

Tables of PVF, CPV and TVF are to be found in Appendix 6 for ascending values of r and n.

8.5   Effects of general inflation and fuel inflation

Case study 8.2
The cost of an energy saving measure is £5000 and the net annual savings are £900:

i   Determine the simple payback in years.
ii   Find the payback period given a discount rate of 10%, a fuel inflation rate of 5% and a general inflation rate of 2%.

Solution
Table 8.3 summarises the solution. The factors in columns 4, 5 and 7 can be calculated using either the formulae or the tables in Appendix 6. You should now confirm the figures and calculations in the table.

Conclusion drawn from Case study 8.2
The discounted payback from Table 8.3 is just over 7 years.

8.6   Net present value and comparison of different schemes

There is frequently a requirement to compare two or more proposals for the purposes of making a decision relating to the most financially viable project. An example of this is found in the following case study.

Table 8.3 Summary of the solution for Case study 8.2

Start of year	Cost of saving (£)	Energy saving (£)	General inflation (2%)	Fuel inflation (5%)	Net energy saving (£)	PV factor (10%)	PV (£)	Cumul. PV (£)
0	−5000	–	–	–	−5000	–	−5000	−5000
1		900	0.98	1.05	926.10	0.909	841.82	−4158.18
2		900	0.961	1.103	953.98	0.826	787.99	−3370.18
3		900	0.942	1.158	981.75	0.751	737.30	−2632.89
4		900	0.924	1.216	1011.23	0.683	690.67	−1942.22
5		900	0.906	1.276	1040.45	0.621	646.12	−1296.10
6		900	0.888	1.340	1070.93	0.564	604.00	−692.10
7		900	0.871	1.407	1102.95	0.513	565.81	−126.29
8		900	0.853	1.478	1134.66	0.467	529.89	+403.60

Case study 8.3

An oil-fired boiler plant has come to the end of its useful life and a comparison is to be made between the installation of a new oil-fired boiler plant and a new gas-fired plant. After flushing out and replacing pumps and controls the existing space heating system is otherwise satisfactory and will have an increased life of at least 20 years which will be the life of the new boiler plant.

As work to the rest of the space heating system and removal of the existing boiler plant is similar for both proposals it does not need consideration in the comparison. The following tabulated information (Table 8.4) relates to each of the two proposals.

Solution

A discount rate of 4% will be applied for Case study 8.3. The factor to be applied for recurring annual expenditure is calculated from Equation 8.2 or from the appropriate table in the Appendix 6. The life of the system is taken as 20 years $(n = 20)$ and $r = 4\%$.

Thus from Appendix 6 or Equation 8.2

$$CPV = PVA = 13.59$$

The cost of repairs at set intervals are single items of expenditure and the factor to be applied is calculated from Equation 8.1 or obtained from the present value table in Appendix 6.

When $n = 3$ PV factor $= 0.889$
$n = 8$ PV factor $= 0.7307$
$n = 12$ PV factor $= 0.6246$
$n = 16$ PV factor $= 0.5339$

Table 8.4 Data for Case study 8.3

Item	Oil-fired plant cost (£)	Gas-fired plant cost (£)	Remarks
Capital costs			
Refurbishment to chimney	1200	4900	Diluted flue for gas
Boiler plant	12 000	9000	
Gas main	–	1600	
Energy costs	6000/yr	4800/yr	
Maintenance and operating costs	1000/yr	800/yr	
Repair costs			
3 years	650	370	
8 years	950	450	
12 years	750	400	
16 years	1000	750	
Repainting oil storage tank and bund £750		–	

Table 8.5 Analysis of oil-fired option, Case study 8.3

Item	Expenditure (£)	Factor	PV (£)
Chimney	1200	–	1200
Plant	12 000	–	12 000
Energy cost	6000/yr	13.59	81 540
Maintenance	1000/yr	13.59	13 590
Repair			
3 years	650	0.889	578
8 years	950	0.7307	694
12 years	750	0.6246	468
16 years	1000	0.5339	534
Repaint			
12 years	750	0.6246	468
		NPV =	111 072

The solution to the oil-fired option is given in Table 8.5.

The solution to the gas-fired option is given in Table 8.6.

Conclusions drawn from Case study 8.3

You will notice here that we are interested in the present cost or net present value. The capital outlay in each proposal is a present cost and therefore has no factor applied. From the tabulated calculations for the two proposals it is clear from the NPV totals that the gas installation is more economical. However, the capital outlay is less for the oil-fired plant: for gas it is £15 500 and for oil it is £13 200.

There is not much difference here in capital outlay but the client would want to know this information as well as the NPV costs.

Table 8.6 Analysis of gas-fired option, Case study 8.3

Item	Expenditure (£)	Factor	PV (£)
Chimney	4900	–	4900
Plant	9000	–	9000
Gas main	1600	–	1600
Energy cost	4800/yr	13.59	65 232
Maintenance	800/yr	13.59	10 872
Repair			
3 years	370	0.889	329
8 years	450	0.7307	329
12 years	400	0.6246	250
16 years	750	0.5339	400
		NPV =	92 912

8.7 Loans

If a building owner has to borrow money to service an energy saving measure the rate of interest chargeable on the loan must be accounted for in the financial appraisal. The interest charged on the loan will be greater than the interest given on a capital investment of the same value. Banks and building societies have rates of interest for investors lower than the borrowing rate in order to pay their employees and make a profit.

In Case study 8.1 the money raised for the energy saving measure was taken from the organisation's reserves which were invested at 5%. If the money had instead been borrowed at an interest rate of 10% the payback period would be extended from 3.33 to 3.74 years. You should now adopt Equation 8.2 and check this solution for yourself.

The same is true in Case study 8.2, thus affecting the calculation of the present value factor PV and the cumulative present value factor CPV.

There are a number of ways in which loan repayments are calculated as shown in Case study 8.4.

Case study 8.4
A loan of £3000 is charged at a borrowing rate of 8% annually over a period of 5 years.

a   Consider a loan the repayments of which are based upon interest on the capital sum

Total repayment $= £3000 + 3000 \times 0.08 \times 5 = £4200$
Annual repayment $= 4200/5 = £840$
Monthly repayment $= 840/12 = £70$

b   Consider a loan the repayments of which are based upon the annual cost method

From the table of CPV factors in Appendix 6 or the use of Equation 8.2, CPV $= 3.993$
The annual cost for repayment of the loan $=$ loan/CPV $= 3000/3.993 = £751$
The monthly premium will be $751/12 = £62.6$
The total cost of the loan will be $= 751 \times 5 = £3755$

c   Consider a loan based upon compound interest

TV $= 3000(1 + 8/100)^5 = £4408$
Annual repayment $= 4408/5 = £882$
Monthly repayment $= 882/12 = £73.5$

Solution and summary

Repayment scheme	Monthly premium (£)	Loan charge (£)
(a)	70	1200
(b)	62.6	755
(c)	73.5	1408

You can see that the repayment or servicing of the loan varies in value and depends upon the method adopted for the calculation.

The annual cost method (b) is normally used for undertaking cost benefit analysis.

8.8 Whole life costs

The cost of owning a system over its life plus the cost in use establishes the whole life cost. The life of various plant and systems is usually between 15 and 30 years. Clearly the life of sophisticated plant will be less than that of benign systems like that of cast iron radiator heating which will extend well beyond 30 years if properly maintained. It therefore may be necessary to consider the whole life of a system in two parts namely plant and distribution.

Whole life costs include the cost of the capital outlay taking account of the interest it could have earned had it been invested or the interest that must be paid on it had it been borrowed. The following case study illustrates whole life costing.

Case study 8.5

The capital cost for a building services installation is £145/m². It is estimated to have a useful life of 20 years. The annual cost in use which includes heating, electricity and maintenance are estimated to be £20/m².

Determine the NPV of the installation and the annual cost of owning and operating it.

If the project is financed out of the organisation's profits or reserves a discount rate of 4% is to be used. If, alternatively, the capital has to be obtained in the form of a loan from a bank, a borrowing rate of 7% is to be adopted. Ignore the effects of inflation.

Solution

We use the table of CPV factors in Appendix 6 or Equation 8.2 for a discount rate of 4%

$$CPV = 13.59$$

The present value of the capital cost $= £145$
The present value of the cost in use $= £20 \times 13.59 = £272$

Total NPV $= £417/m^2$

For a discount rate of 7%,

CPV $= 10.594$

The present value of the capital cost $= £145$
The present value of the cost in use $= £20 \times 10.594 = £212$

Total NPV $= £357/m^2$

To repay a loan of $£145/m^2$ over 20 years at a borrowing rate of 7% using the annual cost method, the CPV factor from the table in Appendix 6 or from adopting Equation 8.2 is 10.594

Loan on first cost payable each year $=$ loan/CPV $= 145/10.594 = £13.69$
Annual cost in use is estimated at £20
Total annual owning and operating cost $= £33.69/m^2$

If the building owner took capital from company profits or reserves to pay for the installation the loss in interest on this capital which would otherwise be invested would be 4% and adopting the annual cost method of repayment the CPV over a term of 20 years will be 13.69

Loan on first cost payable each year $=$ loan/CPV $= 145/13.59 = £10.7$
Annual cost in use is estimated at £20
Total annual owning and operating cost $= £30.7/m^2$

Summary of whole life costs for Case study 8.5
- Net present value of the installation, NPV $= £357/m^2$ borrowed capital.
- Total annual owning and operating cost $= £33.7/m^2$ borrowed capital.
- Net present value of the installation, NPV $= £417/m^2$ investment from capital or reserves.
- Total annual owning and operating cost $= £30.7/m^2$ investment from capital or reserves.
- Note that the NPV of the total annual owning and operating cost of $£33.69/m^2$ on the same terms is $33.69 \times$ CPV factor, which is 33.69×10.594 and NPV $= £357/m^2$.
- Likewise the NPV of $£30.7/m^2 = 30.7 \times 13.59 = £417/m^2$.
- Note also that the NPV is lower for the higher borrowing rate (7%) than it is if investment is taken from capital or reserves (4%).

- The annual owning and operating cost is, however, lower if capital or reserves are used to finance the project.
- Allowances on taxation of company profits may be available on capital expended on investment in the business. This will have the effect of reducing further the total annual owning and operating costs if the organisation finances the project from reserves.
- Another example of whole life costs is the comparison of alternative schemes illustrated in Case study 8.3.

8.9 Repair or replace

When an installation is approaching the end of its economic life, consideration should be given whether to refurbish it or replace it. It is likely that in a number of cases partial replacement will be chosen since parts of the installation may have a longer life, for example the distribution pipework and radiators in a space heating installation.

The factors which require consideration are:

i   Will new plant attract credits from the Climate Change Levy?
ii Cost of replacement
iii Energy saving benefit of new plant
iv Cost in use of new versus existing plant
v   Value placed on reliability and safety if existing plant is kept

Factor (iii) forms part of factor (iv) but is separately listed to emphasise the opportunity that new and more efficient plant will have in saving energy and in reducing carbon dioxide emissions.

The criterion to be followed in making a decision is to compare the NPV of both options. This will include the cost of new replacement plant and its cost in use which will reflect the savings in energy consumption, with the repair costs of existing plant and its cost in use. The analysis can be presented as that in Case study 8.3.

8.10 Chapter closure

You now have some insight into cost benefit analysis which includes simple payback, discounted cash flow and present value, effects of fuel inflation and general inflation, a NPV comparison of alternative schemes, calculation of premiums on loans and life cycle costs. The important focus here is to set out the energy saving proposal or comparison of alternative proposals in such a way for them to be meaningful to the company accountant so that senior management can make financial decisions relating to energy saving proposals alongside other financial matters.

Facility managers will now have to factor in the costs attributed to the Climate Change Levy if no improvement in energy consumption is made.

Chapter 9

Energy audits

Nomenclature

AEC	annual energy consumption (kWh)
AED	annual energy demand (kWh)
LPG	liquid petroleum gas
HWS	hot water services
PI	performance indicator $(kWh/m^2, kg\,CO_2/m^2)$
Benchmarks	$(kWh/m^2, kg\,CO_2/m^2)$
HFI	Heat flux indicator (W/m^2)

9.1 Introduction

An energy audit is now a requirement for non-domestic buildings so that a Building Energy Certificate can be prepared. A definition of the term "energy audit" may help to put this chapter into context.

An energy audit attempts to allocate a value at each point of energy consumption over a given period, usually a year. It should at the least allocate a value to the consumption of the various forms of energy on site.

Following an energy audit it should then be possible to identify at each point of energy consumption the corresponding energy demand. This may require an analysis of energy demand at each point.

The difference between energy consumption and energy demand can then be addressed and ways considered to reduce it at each point to its smallest practical value.

In Chapter 2 the relationship between annual energy consumption, AEC, and annual energy demand, AED, was identified as:

$$AEC = AED/(\text{seasonal efficiency})\,kWh$$

Chapter 11, Section 11.9 introduces the Building Energy Certificate.

9.2   Preliminaries to an energy audit

Before proceeding with an energy audit it is essential that a survey of the staff who use the buildings is undertaken to establish if the comfort levels are acceptable during the winter season and to find out if the provision of hot water for consumption is adequate.

It is not uncommon in a survey of this kind to find that complaints arise due to discomfort from the lack of adequate space heating. This may be due to a variety of reasons. The effect this will have on the energy audit may well be significant as it will show a false annual energy consumption. If the standards of comfort in those areas of complaint are addressed after the audit, there will be an increase in annual energy consumption in the following year. This will give a negative signal to the client or senior management.

If complaints of overheating are found, as evidenced by open windows on cold days, this could be the first issue to be addressed in the programme of energy conservation that follows the audit.

For sedentary occupations in modern well-insulated buildings the majority of people will be neither uncomfortably cool nor uncomfortably warm in winter in rooms where the dry resultant temperature is between 19 and 23 °C.

For active occupations in modern well-insulated buildings the dry resultant temperature should be between 3 and 5 K below that for work of a sedentary nature. Recourse should be made to published data for further information. See Appendix 4.

9.3   Outcomes of the energy audit

Clearly if the efficiency of energy utilisation can be improved, the difference between AED and AEC will be reduced and seasonal efficiency will be increased. Refer to Sections 1.3 and 7.3.

The efficiency of energy use in the case of boiler plant for space heating and the provision of hot water supply over a period will be its seasonal efficiency. Modern boiler plant and controls are more efficient at converting primary energy than older plant. There have been significant advances in boiler and burner design. New plant like the condensing boiler which has a very high efficienciey has entered the market in recent years. The concept of connecting modern low-output conventional boilers in modular format to match changes in load is energy efficient. The move towards generating hot water supply in direct-fired instantaneous heaters is energy efficient. The move towards decentralisation of boiler plant to remove energy losses in distribution is energy efficient although this should not be confused with group heating and district heating schemes, which can make use of refuse incineration, power generation and waste heat.

There are, however, other ways in which annual energy consumption can be reduced and these involve an equally valid approach, namely to reduce the annual demand for energy.

By reducing AED, AEC should also be reduced. There are a variety of action points here and they include:

- Checking maintenance schedules and certificates to ensure that plant and system preventive maintenance has been undertaken. See Appendix 3.
- Checking the time scheduling of plant and zones.
- Checking thermostat settings.
- Checking the thermal insulation on distribution mains.
- Identifying and costing potential energy savings from improvements in the thermal insulation of the building.
- Identifying and costing reduction of the ingress of unnecessary outdoor air into entrance areas and corridors via entrance doors, stairwells and lift wells and around windows.
- Encouraging the occupants of the building to view energy conservation as a daily habit.

The last action point is perhaps the most important. The education of staff in an awareness of energy conservation is paramount to the success of an energy conservation initiative.

The energy audit does not of itself reduce energy use on a site. It can, however, assist in initiating action to reduce the annual energy cost. The energy audit therefore provides two potential courses of action:

- Improving the efficiency of energy utilisation to reduce the difference between AED and AEC to its lowest practical value.
- To reduce annual energy demand to its lowest practical value.

9.4   Measurement of primary energy consumption

In order to prepare an energy audit the invoices for all primary energy supplied to the site must be available for at least the penultimate calendar year. Adjustments may have to be made to ensure that the invoices for different fuels cover the same period. The use of estimated readings should be avoided or checked. At the same time the client should be encouraged to have the various fuels used on site accurately metered so that the fuel accounts for the following year are precise.

Measurement of the consumption of electricity, gas and bulk fuels such as petrol, diesel oil, heating oil, coal, bottled propane and butane can be monitored at the incoming point on site for each. However, the client should be advised to meter the fuels and energy consumption at the various points of use so that in time a more detailed energy audit at points of use can be undertaken.

This type of audit identifies annual energy consumption at points of use and it is from this information that further conservation studies emanate.

Refer to Appendix 5 for details of monitoring equipment.

9.5 Primary energy tariffs

One of the responsibilities of an energy manager is the negotiation of tariffs with the privatised utilities. It is not within the scope of this book to provide advise in this area that has now become quite specialised. Refer to Section 11.13. There is no doubt, however, that negotiation of competative tariffs will have a bearing upon the annual cost of energy, but it is a one-off procedure – at least until the tariffs come up for renegotiation. It is therefore only one of many other potential energy conservation measures. Energy consumption features to take into account when investigating the energy supply markets are:

- Annual base load requirement
- Peak loads and their frequency and points of occurence
- Penalties for exceeding peak loads (and ways this can be avoided by load shedding)
- Seasonal load
- Security of supply (and the potential need for back up supplies)
- In the consumption of electricity, power factor correction.

You can see that these features require detailed knowledge of the way primary energy is consumed on the site. This data may not be fully available from historical records and the energy manager may need to interview key occupants/employees for evidence at points of energy use.

9.6 Presentation of data – A simple audit

The initial annual energy audit can be presented in tabulated format as shown in Table 9.1.

9.6.1 Conversion of energy from fossil fuels to a common base

Energy is normally expressed in MJ or kWh. Table 9.2 below gives conversions for the common fuels.

There now follows an example of an energy audit based upon historical fuel invoices.

Table 9.1 A format for an initial energy audit

1) Fuel	Gas	Oil	Electricity	LPG	Totals
2) Consumption (litres, kWh, bottles)					
3) Consumption in kWh					Total
4) Annual cost (£)					Total
5) Cost/kWh					
6) % Total cost					100
7) % Total consumption					100

Item (1) The table should show all types of energy used on the site; four examples are shown here.
Item (2) Consumption should be based upon fuel invoices and best estimates where necessary.
Item (3) Consumption figures converted to a common unit either GJ or kWh.
Item (4) The total annual cost of the energy/fuel inclusive of standing charges where appropriate.
Item (5) Item (4)/Item (3).
Item (6) Item (4)/total cost in Item (4)
Item (7) Item (3)/total consumption in Item (3).

Table 9.2 Energy conversions for common fuels

Fuel	Energy produced	
	MJ	kWh
1 kg of coal (average)	27.4	7.61
1 litre of petrol	40	11.12
1 cubic metre of natural gas	38.7	10.76
1 cubic foot of natural gas	1.1	0.305
1 cubic metre of propane gas	92.6	25.72
1 kg of propane gas	50	13.9
1 litre of light fuel oil	40.5	11.26
1 litre of heavy fuel oil	41.2	11.44
1 kWh of electricity	3.6	1.0

Case study 9.1

Fuel consumption and cost data are drawn from the previous year's accounts of an organisation and detailed below. Prepare an initial energy audit and comment upon the results.

Gas: 34 771 m³ at £7483, light grade oil: 8000 litres at £1600, electricity: 110 000 kWh at £9900, propane: 1600 kg at £800.

Solution

Clearly much work on the part of the energy manager dealing with this site has been completed already in the production of the fuel account data since it will not be readily available. It is likely that key staff were unaware that the organisation was buying in four different forms of energy.

Table 9.3 Solution to Case study 9.1

Fuel	Gas	Oil	Electricity	Propane	Totals
Consumption	34 785 m³	8000 litres	110 000 kWh	1600 kg	
Consumption (kWh)	374 287	90 080	110 000	22 240	596 607
Annual cost (£)	7483	1600	9900	800	19 783
Cost (p/kWh)	2	1.78	9	3.6	
Percentage of total cost	38	8	50	4	100
Percentage of total consumption	63	15	18	4	100

It is also possible that senior management would not immediately know that it spent £19 783 in the previous year on energy.

The results are given in Table 9.3.

Comments on results of Case study 9.1
- The total cost of energy for the previous year is now known.
- As would be anticipated the major fuel is gas, at 63%.
- The most expensive "fuel" is electricity making up 50% of the annual energy cost at only 18% of annual energy consumed.
- Unless there is obvious potential for saving energy in the use of the other fuels, it is clear that gas and electricity consumption should be investigated first since the major fuel consumption will offer the greatest scope for energy saving and the most expensive fuel will offer the greatest cost benefit for every unit of energy saved.
- At present the cost of oil corresponds to that for gas and if the boiler plant is coming to the end of its working life it may be worth investigating the market for high efficiency dual fuel plant to take advantage of a change in tariffs. However, oil has to be stored on site and this facility may take up valuable space and therefore generate a capital and ongoing cost.
- It is important having produced the energy audit to identify an immediate potential cost benefit which could result from an energy saving measure. This will help to justify the cost of the energy manager's services.

   If the organisation is typical of most and had not previously undertaken an energy audit it is most likely that a low cost energy saving measures like staff awareness and restructuring the fuel tariffs will bring a cost benefit with no capital outlay.
- If the floor area of the premises is calculated, performance indicators can be determined and compared with published benchmarks. See Chapter 6.

9.7   Auditing a primary school

Case study 9.2

The following data was obtained from a primary school in 1995/1996. The school was built in 1986 for a maximum of 240 pupils and is on two storeys with the classrooms around the assembly hall which has clerestory windows for natural light. It is located in Southern England. The plant consists of a single boiler used for space heating and hot water supply to toilets and kitchen.

Floor area is $1027\,m^2$ and the annual consumption of natural gas was $198\,672\,kWh$ at a cost of £2014.85. The annual consumption of electricity was $39\,421\,kWh$ at a cost of £3743.65.

From Appendix 4 the "good practice" benchmark for natural gas is $113\,kWh/m^2$ and $22\,kWh/m^2$ for electricity. The corresponding benchmarks for CO_2 emissions are $21.5\,kg/m^2$ for natural gas and $9.46\,kg/m^2$ for electricity.

Solution

Table 9.4 shows the audit analysis which you should follow through and confirm.

Conclusions and report on school audit

From the audit analysis in Table 9.4 for the primary school the annual target costs and percentage annual savings targets are calculated using the good practice benchmarks, in kWh/m^2.

You should now calculate the annual carbon dioxide emissions for the school in tonnes and compare this with the good practice benchmark emissions.

Table 9.4 Audit analysis for the primary school

Item	Gas	Electricity	Total
Annual kWh	198 672	39 421	238 093
Annual cost (£)	2015	3744	5759
cost (p/kWh)	1.01416	9.4966	
% of total cost	35	65	100
% of total consumption	83	17	100
Performance indicator (kWh/m²)	193	38	231
Benchmark (kWh/m²)	113	22	135
CO₂ Performance indicator (kg/m²)	38.6	26.6	65.2
Benchmark (kg CO₂/m²)	21.5	9.46	31
Annual target cost (£)	1180	2167	3347
Annual savings target (£)	835	1577	2412
% Annual savings target	41.4	42.1	41.8 (average)
Audited cost per pupil (£)	8.40	15.6	24

Clearly there are savings to be made here and in any case the performance indicators for the school are well above the benchmarks. It will, however, be difficult to reach the annual target savings of £2412 without replacing the plant and controls.

During the site visit prior to preparing the audit it became evident that savings could be made. This is often the case. Here are some pointers.

- The old style 20 mm diameter fluorescent tubes were in use throughout the classrooms. Each of these could be replaced as they end their working life.
- The space heating system at that time was 10 years old and the single cast iron boiler provided both the space heating and hot water supply.
- The boiler did not have a bolt on cycling management system.
- There was no one responsible for the space heating and electrical systems at the school.
- Annual maintenance was not properly supervised with little feedback to the head teacher.

Governors report

A number of questions need to be addressed to the governing body and in turn to the staff bearing in mind that the purpose of an audit and its conclusions is to present lay people with information and recommendations they understand. It is important that staff, caretaker and governors are enabled to own the audit. This task lies with the auditor.

Hopefully with help they will see and understand the significance of the analysis in Table 9.4. Here are some pointers.

- Is the school a comfortable building to work in during the winter and when natural light levels are low? Is additional heating and/or lighting required? Positive responses here may inflate the cost of the fuel account.
- How do staff respond to energy conservation in the school on a daily basis relating to the use of lighting and space heating?
- Is energy conservation at the school embedded in the curriculum and do the children participate?
- Is the head teacher aware of the time scheduling of the plant?
- How is the service contract supervised?
- Where do staff feel energy savings can be made, taking the analysis in Table 9.4 into account.

The next case study relates to a small museum specialising in local history of city development in the 18th century.

9.8   An energy audit for a small museum

Case study 9.3

The audit was undertaken in 1996/1997. The building is listed grade two and was built two hundred years ago. It consists of a large hall in which the artifacts are located, an unheated balcony and adjoining offices and toilets. The floor-to-ceiling height of the hall including the balcony is 10 metres.

The building is of stone ashlar, single skin and unlined on the inside surface with tall, arched, single, glazed windows on two sides. the floor is of stone flags that were re-laid at the same time as the heating system was installed during its earlier refurbishment. The roof is a pitched structure with a layer of thermal insulation between the joists. There is a curator who divides her time between the museum and other historic buildings in the city. When the museum is open to the public it is staffed by volunteers and visitors to the museum are transient.

The space heating system consists of an atmospheric gas–fired boiler serving cast iron radiators and pipe coils and was probably installed 10 years before the audit. Electric instantaneous water heaters are provided in the toilets which are remote from the boiler. Floor area is $284\,m^2$ and the energy invoices total $76\,284\,kWh$ for natural gas and $30\,759\,kWh$ for electricity. The annual cost for gas was £976.43 and for electricity £3091.3.

Solution

Table 9.5 sets out the audit analysis. Good practice benchmarks for energy are taken from Appendix 4 and for carbon dioxide emissions from Table 6.1.

You should now follow through and confirm the solutions in the table.

Table 9.5 Audit analysis for the museum

Item	Gas	Electricity	Total
AEC (kWh)	76 284	30 759	107 043
Annual cost (£)	976.43	3091.3	4067.73
p/kWh	1.28	10.05	
% Cost	24	76	100
% Consumption	71.26	28.74	100
Performance indicator (kWh/m²)	268.6	108.3	376.9
Benchmarks (kWh/m²)	96	57	153
Performance indicator ($CO_2\,kg/m^2$)	51.03	46.57	97.6
Benchmarks ($kg\,CO_2/m^2$)	18.24	24.51	42.75
Target costs (£)	349	1672	1976
Annual savings target (£)	627	1464	2091
% Target savings	35.7	52.6	48.6 (average)
Annual cost (£/m²)	3.44	10.9	14.34
Target cost (£/m²)	1.23	5.73	6.96

Conclusions and report on the museum audit

- The annual emission of carbon dioxide in 1996/1997 is 28 tonnes compared with the good practice benchmark of 12 tonnes.
- Clearly there are savings to be made here with such a wide disparity between the performance indicators and the good practice benchmarks. Account must be made of the age of the building and the fact that it is listed. However, it is likely that significant reductions in energy consumption can be made without substantial expenditure.
- The time scheduling of the plant is unclear and needs investigation. It is likely that savings can be made here by matching the preheat and plant run times more accurately with the opening times.
- More roof insulation would be a serious energy saving measure. Cost, savings and length of payback can be calculated as shown in Example 2.1.
- Lining the inside of the hall would be a major undertaking but it would improve the thermal transmittance through the external walls and reduce the preheat times and of course save energy. Again costs, savings and payback can be calculated.
- Lighting in a museum is an important feature but savings can be made using low voltage luminaires for the displays that are at present lit by 230 volt spot lamps.
- One of the offices is heated by an on-peak electric convector. It would seem sensible to investigate the installation of a radiator connected to the space heating system if the time scheduling coincides with that of the museum.

9.9   Auditing a mixed use building

The next case study is in two parts. The first looks at an initial audit evaluated from invoices for gas meter and electricity meter readings for a site. The second audit is determined from more detailed observations on the same site for the same calendar year. It will show the value of investing in metering equipment.

Case study 9.4
A building consisting of five storeys and having a treated floor area of $2000\,m^2$ is divided into offices, restaurant, gymnasium and aerobics room. The quarterly gas and monthly electricity consumptions were taken from the main utility meters and are given in Table 9.6.

Solution
You will notice that the electricity readings were taken monthly by staff but the gas consumptions were based upon the quarterly readings from a gas meter registering in hundreds of cubic feet which have been adjusted

Table 9.6 Initial gas and electricity consumptions for Case study 9.4

Month	Gas (ft³)	Electricity (kWh)	Month	Gas (ft³)	Electricity (kWh)
September	–	5320	March	–	7000
October	–	6450	April	–	6375
November	233 364	6675	May	419 327	5815
December	–	6850	June	–	4140
January	–	7305	July	–	4320
February	587 058	7380	August	27 347	4370

Table 9.7 Analysis of the initial energy audit for Case study 9.4

Fuel	Gas	Electricity	Totals
Annual consumption	1 267 096 ft³	72 000 kWh	
Consumption (kWh)	386 464	72 000	458 464
Annual cost (£)	5024	4320	9344
Cost (p/kWh)	1.3	6.0	
Percentage of total cost	54	46	
Percentage of total consumption	84	16	

accordingly. Quarterly charges have been omitted to aid clarity. Clearly the initial audit will yield a limited analysis of energy consumption for this site. Table 9.7 shows the analysis.

Table 9.7 is set out and the figures calculated in a similar manner to Table 9.1 along with the associated notes.

Comments on the initial analysis of Case study 9.4

The comments which can be made here are limited to the annual total cost of energy at £9344 and the rather obvious conclusion that although electricity is only 16% of the total energy consumption it represents 46% of the annual energy cost.

It would be far more beneficial if the ways in which gas and electricity were consumed were broken down so that a more detailed analysis can be undertaken. This will however require the investment in the purchase (or hire) and installation of metering equipment.

9.10   Presentation of data – A more detailed audit

There now follows the second part of the case study (Case study 9.5) in which metering equipment was installed so that the following services could be analysed:

- Space heating to radiator circuits and the air handling unit serving the warm air system.
- Hot water supply to the restaurant, showers and wash hand basins in the toilets.
- Electricity supplies to the lighting circuits, lifts, mechanical services plant and small power.

Appendix 5 contains a list of metering and monitoring equipment currently available on the market.

Case study 9.5

The observations which were taken during the course of the year are given in Tables 9.8 and 9.9. The treated floor area for catering was found to be $25\,m^2$. On investigation it was found that the hot water supply was generated from centralised instantaneous gas direct fired heaters while the space heating system had its own dedicated plant consisting of two boilers in parallel operating on sequence control.

The radiator system accounts for the structural losses in the building and the air handling unit provides tempered air to the building via supply and return ductwork thus accounting for the heat losses resulting from the rate of air change. Table 9.8 shows the record of observations of heat energy and electricity consumptions. Table 9.9 shows the record of observations of gas consumption for space heating and hot water supply.

Solution to Case study 9.5

The data collected from the additional metering equipment installed in the building are given in Tables 9.8 and 9.9. You will see there are two further gas meters each of which has independently recorded the gas consumption for the heating and hot water supply. There are five heat meters installed to independently record the heat energy consumption for the radiator system, the warm air system via the air heater battery and the hot water supply to the restaurant, showers and wash hand basins.

There are also subsidiary electricity meters for independently recording the electricity consumption for the lighting, lifts, power requirements and services plant.

Clearly this will produce an energy audit with much more detailed analysis. Table 9.10 gives the analysis of the observations recorded in Table 9.8.

Comments on the analysis in Table 9.10

The Annual Energy Consumptions in Table 9.10 are those consumed at the points of use. They are not Annual Energy Demand (AED) totals which do not account for the energy utilization and hence seasonal efficiencies of plant and systems.

Table 9.8 Annual record of observations for Case study 9.5

Month	Heat energy consumption (kWh)				Electricity consumption (kWh)				
	Rads	AHU	Catering	Showers	Basins	Lighting	Lifts	Power	Services plant
September	–	–	1389	556	278	1800	620	1800	1100
October	2778	1944	1806	972	611	2020	630	2500	1300
November	22 222	11 111	1806	972	611	2200	625	2450	1400
December	25 000	12 500	1806	972	611	2250	625	2475	1500
January	23 611	11 111	1806	972	611	2850	630	2475	1350
February	27 778	13 889	1806	972	611	2860	620	2500	1400
March	23 611	11 111	1806	972	611	2500	625	2475	1400
April	19 444	9722	1806	972	611	2000	625	2450	1300
May	5556	4167	1806	972	611	1700	640	2475	1000
June	–	–	1389	556	278	820	620	1900	800
July	–	–	1389	556	278	700	620	1800	1200
August	–	–	1389	556	278	800	620	1700	1250
Totals	150 000	75 555	20 004	10 000	6000	22 500	7500	27 000	15 000
Grand totals	261 599 kWh				72 000 kWh				

Table 9.9 Gas meter readings for Case study 9.5

Month	Gas meter readings (ft³)	
	Heating	Hot water supply
September	–	9116
October	25525	14585
November	169554	14585
December	173200	14585
January	171377	14585
February	198724	14585
March	175935	14585
April	148588	14585
May	51048	14585
June	–	9116
July	–	9116
August	–	9116
Totals	1113951	153144
Grand total	1267095 ft³ (compares with Table 9.7)	

Table 9.10 Common annual energy consumption totals from the record of observations in Table 9.8 Case study 9.5

Fuel service	Gas heating		Gas HWS		Electricity				
	Rad system	AHU	Catering	Showers	Basins	Lighting	Lifts	Power	Services plant
Annual kWh	150000	75555	20004	10000	6000	22500	7500	27000	15000
Grand totals:	Heating 225559 kWh		HWS 36000 kWh			Electricity 72000 kWh			

The Annual Energy Consumption (AEC) for natural gas is 261599 kWh and the AEC for electricity is 72000 kWh.

Seasonal efficiencies of the plant and systems

If the energy consumption figures for gas in Table 9.9 are considered, the gross values of energy input of gas in Table 9.7 of 1267095 ft³ or 386464 kWh can be split into that required for the space heating and that needed for the hot water supply. In Table 9.10 the net grand totals of heat energy used for space heating and hot water supply are given. In Table 9.9 the gross grand totals of gas consumed are given. The seasonal efficiencies can therefore be determined for the heating and the HWS from this data and these are shown in Table 9.11.

You will see that the total consumption of natural gas in Table 9.11 comes from Table 9.9. The so-called points of use consumptions in kWh are the annual energy consumption totals obtained from Table 9.10.

Table 9.11 Seasonal efficiency of space heating and HWS systems
Case study 9.5

Item	Space heating	Hot water supply
Gas consumption (ft³)	1 113 951	153 144 (from Table 9.9)
Conversion to kWh	339 755	46 709 (Table 9.2)
Points of use (kWh)	225 559	36 000 (from Table 9.10)
Seasonal efficiency (%)	66	77

Table 9.12 Performance Indicators for the services identified in Case study 9.5

Service	Net consumption (kWh)	Seasonal efficiency (%)	Gross consumption (kWh)	Treated area (m²)	PI (kwh/m²)
Rad system	150 000	66	227 273	2000	114
AHU	75 555	66	114 479	2000	57
Catering	20 004	77	25 974	25	(1039)
Showers	10 000	77	12 987	2000	6.5
Basins	6000	77	7792	2000	3.9
Lighting	22 500	–	22 500	2000	11.25
Lifts	7500	–	7500	2000	3.75
Power	27 000	–	27 000	2000	13.5
Services plant	15 000	–	15 000	2000	7.5
Total			460 505		

Total PI excluding catering 217 kWh/m²
0.78 GJ/m²

It can be seen from Table 9.11 that the seasonal efficiency for the hot water supply system is high. This is to be expected for direct fired instantaneous HWS plant. These seasonal efficiencies are not to be confused with the manufacturers' efficiencies of plant items only that are done under ideal test conditions and do not account for use over time or the systems they are connected into.

Determination of Performance Indicators
A series of performance indicators can be generated from the data in Tables 9.10 and 9.11.

The treated floor area for the building is given as 2000 m² and if the seasonal efficiencies from Table 9.11 are used for each point of use for the heating and hot water supply systems the Performance Indicators for the services can be determined as shown in Table 9.12.

Comments on the Analysis of Case study 9.5 and Table 9.12
• The performance indicator for the showers is derived from the total treated floor area inferring that the gymnasium and aerobics rooms are used by the occupants of the whole building which may not be

the case. Further investigation into the use of these areas would be helpful.

- The performance indicator for the catering services is for the consumption of natural gas, it does not include the electrical power consumption.
- The PI for catering can be compared with the Benchmarks given in Table 9.13 and taken from the Energy Efficiency booklets: Introduction to Energy Efficiency in catering establishments. See Appendix 4.
- You should now refer to Chapter 6 in which energy targets expressed in kWh/m^2 for building services are given from the Building Analysis files in the CIBSE monthly Journals. These can be compared with the results in Table 9.12.
- The gross energy consumptions calculated at each point of use in Table 9.12 now allow the determination of the percentage of total site consumption of energy for each point of use.
- The percentage of energy consumption for the points of use for heating and HWS can also be determined and using the annual cost for these services, the cost of each service at the point of use can be calculated. Refer to Table 9.14.
- A similar calculation can be done for the electrical services. This analysis is also set out in Table 9.14.

Table 9.13 Benchmarks for catering establishments

Restaurant with bar	1100–1250 kWh/m²
Fast food restaurant	480–670 kWh/m²
Pub restaurant	2700–3500 kWh/m²

Table 9.14 Cost analysis of data from Tables 9.7 and 9.12 Case study 9.5

Point of use	% Site consumption	% Totals	Total costs (£)	% Heating and HWS consumption	Cost at point of use (£)
Radiators	49.4			58.5	2939
AHU	24.9			29.4	1477
Catering	5.6			6.7	337
Showers	2.8			3.3	166
Basins	1.7			2.0	100
Sub totals		83.4	5024	100	5024
				% Electrical consumption	
Lighting	4.9			31.3	1352
Lifts	1.6			10.4	449
Power	5.9			37.5	1620
Services plant	3.3			20.8	899
Sub totals	100	15.6	4320	100	4320
Totals					9344

With the completion of Table 9.14 a fairly detailed picture of energy use on this site now emerges. The audit has provided the annual energy consumption and cost at each point of use as well as the performance indicators for each of the services in the building.

Action points

It is left to you to prepare a summary and action points for improving the energy performance of the services in this building. These need to be included with the tables of audit for the client. Have a look initially at the comments made for Case study 9.1. Also refer to Appendices 2 and 3.

Clearly the resolution of the data in Tables 9.7, 9.8, 9.9, 9.10, 9.11, 9.12, 9.13 and 9.14 can be generated using a spread sheet or data base to take out the tiresome long hand calculations.

Qualifying remarks

Qualifying remarks in the report should include:

- There will be margins of error in the sub-metering equipment used to determine the seasonal efficiencies for the space heating plant and the plant generating the hot water supply.
- The demand for electricity is assumed to be equal to the consumption which implies a conversion efficiency of 100%. This does not identify the operating efficiencies of pumps, fans, lifts etc.
- It is now left to you to identify further qualifying remarks for this audit.

This completes the analysis for Case study 9.5.

9.11   Auditing a two-bed bungalow Case study 9.6

A detatched bungalow located in the Severn Valley has a floor area of $100\,m^2$ and is occupied by two adults. It is audited in Table 9.15. The audit includes gas heating and hot water supply, electricity use for power, lighting and cooking and petrol consumption from the use of a car.

Table 9.15 Audit details for Case study 9.6

Item	Gas	Electricity	Petrol	Totals
Annual kWh	19278	5220	13944	38442
Annual kWh target	7711	2088	5578	15377
Annual kg CO_2	3663	2245	3765	9673
Annual kg CO_2 target	1465	898	1506	3869
Performance Indicator kWh/m^2	193	52	–	245
Performance Indicator kg CO_2/m^2	36.7	22.4	–	59.1
Benchmark target kWh/m^2	77.11	20.88		98
Benchmark target kg CO_2/m^2	14.65	8.98		23.63

The car covered an annual distance of 7840 miles that converts to 12 544 km, petrol consumption being 10 km/litre. Table 9.2 suggests 1 litre of fuel is equivalent to 11.12 kWh. Table 6.1 gives the conversion factors for $kg\,CO_2/kWh$. Included in the audit are annual targets in kWh and $kg\,CO_2$ for a 60% reduction by the year 2050 (the 40% house) as set out in the Government's White Paper of 2003.

Solution
Table 9.15 sets out the audit details.
You should now confirm agreement of the solutions in Table 9.15.

Conclusions
- As the climate gets warmer CO_2 emissions should fall.
- If the Gulf Stream fails as some scientists predict the climate in the UK will get colder and CO_2 emissions will rise.
- Current annual CO_2 emissions in this case study are 4.84 tonnes/person.
- Target annual CO_2 emissions by 2050 are 1.94 tonnes/person.
- Annual CO_2 emissions in this case study exclude that generated by water supply, sewage treatment, refuse collection and domestic waste management.
- Annual CO_2 emissions in the UK for the year 2003 were 560 million tones. This is equivalent to 9 tonnes per person. The difference accounts for the industrial, commercial and transport emissions. Nevertheless 4.84 tonnes of domestic emissions per person is 50% of the total emissions per person and underlines the need for domestic reduction in the use of fossil fuels.
- A report in the May CIBSE Journal of 1999 gave a highly insulated newly built house an annual energy consumption of $120\,kWh/m^2$. This compares with the year 2050 benchmark of $98\,kWh/m^2$. Whether existing housing stock can be reduced to this level of annual energy consumption is a question you might like to consider baring in mind that there is a direct correlation between energy consumption from fossil fuels and carbon dioxide emissions. Refer to Section 3.9. Section 11.11 introduces current legislation.

9.12   Further source material

Recourse can be made to the Energy Efficiency Office's Fuel Efficiency Booklets and CIBSE Guide book F.

9.13   Chapter closure

You now have a clear idea of the place of the energy audit in the preparation of a Building Energy Certificate, what an energy audit is, how to prepare it and the courses of action that can be taken following the audit. The measurement of the consumption of energy on the site is set out and the discussion on primary energy tariffs should assist in identifying the factors for consideration when negotiating with the energy suppliers.

You are now aware how the audit can be extended to the point of use by employing sub-metering equipment.

Chapter 10

Monitoring and targeting

Nomenclature

LTHW	low temperature hot water heating
HTHW	high temperature hot water heating
LPG	liquid petroleum gas
r	correlation coefficient
DD	Degree Days (K·days)
x	independent variable
y	dependent variable
a, b	regression coefficients
n	number of observations under revue
t_m	24 hour mean daily temperature (°C)
d	temperature rise due to indoor heat gains (K)
N	number of days in period under review
SDD	Standard Degree Days (K·days)
t_b	Base temperature/Control temperature/Balance temperature
t_i	indoor design temperature (°C)
CUSUM	cumulative sum deviation
HWS	hot water services

10.1 Introduction

When a building owner has had an energy audit completed for the previous year it is very important to follow this up with an energy agenda to ensure that interest in the management of energy is focused and does not peter out.

The audit in addition to setting out the annual costs of energy consumptions on site can promote a focus in two areas of energy conservation:

- Reduction in the difference between annual energy demand and annual energy consumption. This can be achieved by using efficient plant and

equipment and ensuring that distribution pipework is efficiently insulated and by utilising energy efficiently.
- Reduction in the demand for energy to its lowest practical value. This is addressed by utilising energy efficiently at each point of use.

The tariffs for primary energy might be considered a third focus. However, this issue should be addressed immediately following the audit as a matter of course if it has not been done before, and suggestions on factors to consider in negotiating a tariff are included in Section 9.5.

Energy conservation is considered in Chapter 7. The purpose of this chapter is to consider how energy consumption can be monitored and the ways in which checks can be made to verify that energy saving measures are yielding the estimated benefits of lower energy consumption from fossil fuels.

In other words we need to show that a building's performance indicator is improving and moving towards the good practice benchmark or other target set for the building. The process of monitoring will also identify whether the consumption of energy is following the expected trend.

10.2 Monitoring procedures

Monitoring the consumption of energy should follow an initial energy audit that is based upon a previous year's fuel invoices. It will meet three main objectives:

- It will provide a more detailed annual audit.
- It will provide data for more detailed analysis at points of energy use to establish patterns and variations.
- It will provide data for a system of continuous performance monitoring.

These objectives will assist in having the site's energy consumption under continuous scrutiny.

The first step in monitoring is to prepare a block diagram. This identifies the locations of energy input, the plant that converts it for use, the media where appropriate for transportation to site locations and the services provided at the point of use. Figure 10.1 is a typical block diagram for site space heating, hot water supply, lighting and power.

The energy audit will have established the energy inputs. The block diagram should identify energy conversion plant and prime movers, transport media and distribution, and the services provision. In the process its preparation should give an insight to the extent and size of the systems that directly or indirectly use fossil fuels on the site.

Energy input	Conversion plant	Transport media	Service
Gas	Boilers	LTHW/HTHW	Space heating
Oil	Air handling plant	HWS secondary	HWS
Electricity	Pumps	Electricity	Lighting
Solid fuel	Fans	Air	Air heating
Ventilation	Controls	Steam	Power
LPG			

Figure 10.1 Block diagram for site space heating, hws, lighting and power.

10.3 Monitoring equipment

Having identified the points of energy input on the site, an investment is usually required for monitoring and recording equipment to be located at the points of energy use such as gas, oil, LPG and electricity. Most of this equipment can be hired for use. It includes multi-channel data loggers, check meters, hours run meters, oil/water flow meters, gas meters, temperature sensors and a flue gas analyser for combustion efficiency tests. You should now refer to Appendix 5 for a more extensive list. The frequency and accuracy of recording consumption data will depend upon the level of commitment. The reading and recording of the main meters and fuel deliveries must be a priority and should not be left to the fuel suppliers. Readings should be taken monthly, if not weekly, preferably on the same day and at the same time. A data record form should be devised for this purpose.

10.4 Correlation and linear regression analysis

Performance monitoring involves taking a number of pairs of readings for the purpose and plotting them on a graph to generate a thermal performance line that visualises system performance.

Two common pairs of readings used to measure the performance of space heating systems are:

- Energy/fuel consumption and Degree Days.
- Energy/fuel consumption and average mean daily outdoor temperature.

The correlation coefficient can have two functions:

- It establishes whether or not there is a degree of association between two variables.
- It validates the accuracy of the observations.

Table 10.1 Minimum correlation coefficients for acceptable levels of association. The correlation coefficients will have positive or negative values

Number of observations	Minimum correlation coefficient
10	+/ − 0.767
15	+/ − 0.641
20	+/ − 0.561
25	+/ − 0.506
30	+/ − 0.464
35	+/ − 0.425
40	+/ − 0.402

Clearly observations of fuel consumption and mean outdoor temperature for a building will have a strong degree of association. It is therefore the second function of the correlation coefficient that is needed to validate the accuracy of the observations taken.

If the calculated coefficient is below the minimum value given in Table 10.1, the observations are not valid, which is to say they are not sufficiently accurate. This means that more care must be taken in the process of recording fuel consumption and mean daily outdoor temperature.

Correlation is the degree of association between two unrelated quantities which are varying together. The correlation coefficient r is the measure of the association and varies between −0.99 and +0.99. A significant correlation is established if the coefficient is in excess of a given value that will vary with the number of events or observations in the sample. Table 10.1 lists minimum correlation coefficients for acceptable levels of association between a dependent and an independent variable against the number of observations.

You will see that as the number of events increase so the value of the minimum correlation coefficient reduces.

In research work any two variables can therefore be compared and the correlation coefficient calculated to see if there is an association between them. It is apparent that there should be a significant correlation between the pairs of observations suggested above. This will be tested in the following analysis. When they are plotted on a graph it is likely that there will be a scatter of points and hence some difficulty in placing the line of best fit. It is at this juncture that the adoption of another mathematical tool will be useful.

The following case study introduces regression analysis using monthly energy consumption and Degree Days.

Case study 10.1
Consider the data in Table 10.2 relating to monthly energy consumption by a space heating plant and the monthly Degree Days for the locality.

Table 10.2 Monthly Degree Days and energy consumption in GJ for a consumer

Month	J	F	M	A	M	J	J	A	S	O	N	D	Total
DD	340	370	280	230	160	40	–	–	–	110	240	315	2085
Energy in GJ	330	380	310	240	170	150	100	100	100	180	340	270	2670

Solution

Energy here is measured in GJ that can easily be converted to kWh since $1\,GJ = 278\,kWh$.

You will notice from Table 10.2 that in the summer months of July, August and September there is a constant load of 100 GJ. It is clear that this load extends throughout the year and represents the requirement for HWS which is unrelated to the monthly SDD. This is known as the "base load".

If these pairs of readings are plotted on a graph of Degree Days against energy consumption, a scatter of points will emerge as shown in Figure 10.2 and difficulty is experienced in plotting the line of best fit.

The solution involves finding an equation that describes the correlation between the monthly Degree Days, x, and the monthly energy consumption, y. Degree Days x is the independent variable and that part of the energy consumption y which is weather related is therefore dependent upon the Degree Days and is the dependent variable. The equation of regression line is thus y on x. The technique of finding the equation is known as "regression analysis". Linear regression which applies here implies a straight-line association between x and y and the equation will therefore be of the form

$$y = ax + b \tag{10.1}$$

where a and b are the regression coefficients.

Values of a and b may be found from a pair of simultaneous equations:

$$\Sigma y = a\Sigma x + nb \tag{10.2}$$

$$\Sigma xy = a\Sigma x^2 + b\Sigma x \tag{10.3}$$

where n = number of events or observations.

Tabulated values of x and y are shown in Table 10.3.

The two simultaneous Equations 10.2 and 10.3 now become:

$$2670 = 2085a + 12b$$

and

$$614450 = 579925a + 2085b$$

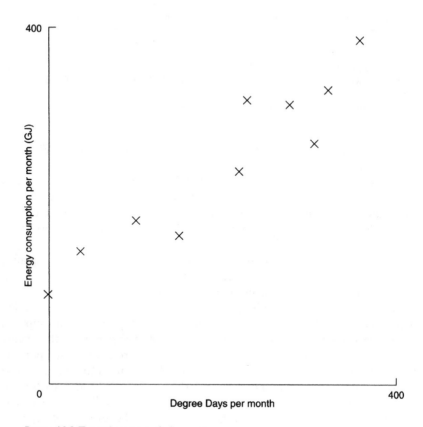

Figure 10.2 Typical scatter of observations.

Table 10.3 Tabulated data based on information in Table 10.2

x	y	xy	x^2	y^2
340	330	112 200	115 600	108 900
370	380	140 600	136 900	144 400
280	310	86 800	78 400	96 100
230	240	55 200	52 900	57 600
160	170	27 200	25 600	28 900
40	150	6000	1600	22 500
–	100	–	–	10 000
–	100	–	–	10 000
–	100	–	–	10 000
110	180	19 800	12 100	32 400
240	340	81 600	57 600	115 600
315	270	85 050	99 225	72 900
$\Sigma x = 2085$	$\Sigma y = 2670$	$\Sigma xy = 614 450$	$\Sigma x^2 = 579 925$	$\Sigma y^2 = 709 300$

Multiplying the first equation by 174

$$464\,580 = 362\,790a + 2085b$$

subtracting the third equation from the second:

$$149\,870 = 217\,135a$$

from which

$$a = 0.69$$

Substituting for a in the first equation:

$$b = 103$$

The regression equation 10.1 is therefore:

$$y = 0.69x + 103$$

Two arbitary values can now be given to x within the range from 0 to 400 and y evaluated. Let $x = 100$ then $y = 172$ and if $x = 350$ then $y = 345$. The two coordinates on the graph can then be connected by a straight line which will be the line of best fit in the scatter of points resulting from the observations. See Figure 10.3. You will see that regression coefficient b is evaluated as 103. In fact of course it should be 100 which is the energy consumption for HWS. The error occurs in rounding the numbers in the simultaneous equations. As the consumption of HWS is consistent and continuous throughout the year it is called the "base load" which is weather unrelated. In practice the base load consumption of 100 GJ would not be exactly the same each month for HWS. There would also be a shutdown periods for holidays which are not considered here.

It is now necessary to determine the correlation coefficient and comparing with the observations in Table 10.1.

Clearly one would expect a close correlation between energy consumption and Degree Days. The second function of the coefficient therefore applies here in validating the accuracy of the observations in Table 10.2. The value of the minimum correlation coefficient is dependent upon the number of observations. It is also dependent here on the association of energy consumption and Degree Days. The energy consumption for HWS is therefore unrelated. However, if it is consistent throughout the year it is only necessary to omit the summer months of July, August and September when no heating is required. Table 10.3 must therefore be adjusted by omitting the

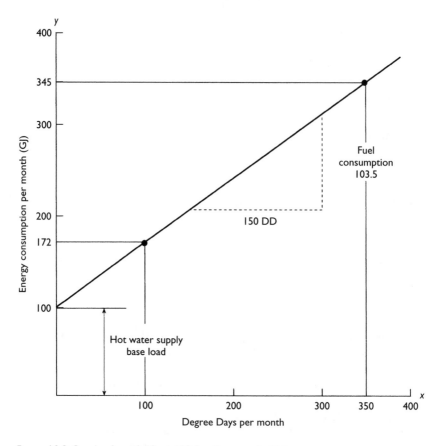

Figure 10.3 Graph of $y = 0.69x + 103$ for Case study 10.1.

three summer months making the number of observations 9 instead of 12. This will affect the summations of y and y^2 in Table 10.3.

The formula for the calculation of the coefficient r is:

$$r = (\Sigma xy - nxy)/\sqrt{(\Sigma x^2 - nx^2)(\Sigma y^2 - ny^2)} \qquad (10.4)$$

where x and y are the mean values and $n = 9$.

From Table 10.3 $\Sigma y = 2670 - 300 = 2370$ and $\Sigma y^2 = 709\,300 - 30\,000 = 679\,300$.

Substituting these values and the other summations from Table 10.3 into Equation 10.4:

Thus $r = 614\,450 - (9 \times 231.66 \times 263.33)/$

$$\sqrt{(579\,925 - 483\,025)(679\,300 - 624\,100)}$$

$$= 65\,401/\sqrt{(96\,900 \times 55\,200)}$$
$$= 65\,401/73\,136$$
$$= +0.894$$

From Table 10.1, which identifies minimum values for the correlation coefficient, 10 observations has a minimum value for r as 0.767. It is therefore apparent that the observations given in Table 10.2 are validated and therefore accurate. In the determination of the correlation coefficient you may feel unhappy about including the HWS base load in the nine heating months since it is unrelated to Degree Days. Table 10.4 discounts the HWS base load and the subsequent calculation for r you will see still agrees with its evaluation above.

Adopting Equation 10.4 for the correlation coefficient r

$$r = 405\,950 - (9 \times 231.67 \times 163.33)/$$
$$\sqrt{((579\,925 - 9 \times 53\,671)(295\,300 - 9 \times 26\,677))}$$
$$= 65\,400/\sqrt{(96\,900 \times 55\,200)}$$
$$= 65\,400/73\,136$$

Correlation coefficient $r = +0.894$

Remember when two variables have an obvious association the second function of the correlation coefficient applies. If the observations were not recorded properly the correlation coefficient would draw attention to the fact by being below the acceptable minimum value given in Table 10.1. Care should therefore be taken in recording observations in a consistent manner and ensuring that, for example, accurate energy consumption readings are

Table 10.4 Tabulated data excluding the base load

x	y	xy	x^2	y^2
340	230	78 200	115 600	52 900
370	280	103 600	136 900	78 400
280	210	58 800	78 400	44 100
230	140	32 200	52 900	19 600
160	70	11 200	25 600	4900
40	50	2000	1600	2500
110	80	8800	12 100	6400
240	240	57 600	57 600	57 600
315	170	53 550	99 225	28 900
$\Sigma x = 2085$	$\Sigma y = 1470$	$\Sigma xy = 405\,950$	$\Sigma x^2 = 579\,925$	$\Sigma y^2 = 295\,300$

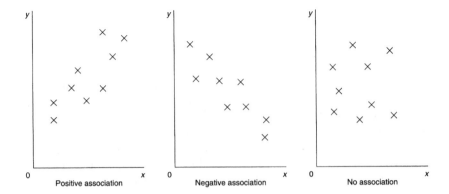

Figure 10.4 Scatter diagrams.

taken at the beginning or end of each calendar month to coincide with the monthly Degree Day totals.

The graphical effects of positive association, negative association and no association are shown on scatter diagrams in Figure 10.4.

Summary of performance analysis, Case study 10.1
- Slope of performance line is +0.69 GJ/DD
- Thermal energy intercept is 103 GJ (100 GJ)
- Correlation coefficient is +0.89, which implies a positive association. See Figure 10.5.

Conclusions
Regression analysis is a key mathematical tool for locating the line of best fit from a number of observations of two related variables in a programme of monitoring energy consumption.

Where the degree of association between two related variables is not immediately apparent, determination of the correlation coefficient will provide the necessary evidence of association. It will also validate the accuracy of the observations.

Regression analysis and correlation can be performed rapidly on a computer spreadsheet or database and on some pocket calculators.

In the drive for savings in energy the energy manager may want to investigate the possibility of a correlation between what at first sight might appear as two unrelated variables to establish if there is a degree of association between them. The following are potential examples:

- Water consumption and mean outdoor temperature
- Power consumption and mean outdoor temperature
- Artificial lighting and levels of daylight.

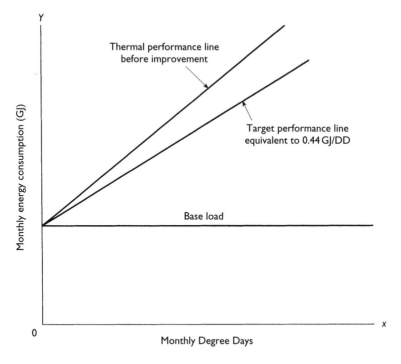

Figure 10.5 Target performance line.

10.5   Continuous performance monitoring using Degree Days

Case study 10.1 is an exercise in the historical performance of energy use for space heating and hot water supply, say, for the previous year. The monthly performance of the space heating system over the 9 months of the heating season can be obtained by dividing the GJ consumed by the Degree Days on a monthly basis as shown in Table 10.5.

The month of heaviest energy use was in June perhaps indicating that the plant was in need of servicing or it may have been cycling as a result of milder weather and hence low load. If the latter was the case, consideration

Table 10.5 GJ/DD for Case study 10.1

Month	J	F	M	Ap	Ma	Jn	O	N	D	Total
DD	340	370	280	230	160	40	110	240	315	2085
GJ	230	280	210	140	70	50	80	240	170	1470
GJ/DD	0.96	0.76	0.75	0.61	0.44	1.25	0.73	1.0	0.54	0.705 (average)

could be given to introducing controls which inhibit boiler plant cycling under these conditions. November was also a poor month but May at 0.44 GJ/DD provides the most efficient use of energy. The monthly variations in GJ/DD over the season's observations would give cause for concern and an investigation might be in order. The plant and system should be able to provide a monthly GJ/DD ratio approaching the best historic value of 0.44. If this is achieved in the following year the slope of the performance line in Figure 10.3 would change.

For monitoring the continuous performance of the space heating system a target performance line could be drawn at a slope equivalent to 0.44 GJ/DD for the following year as shown in Figure 10.5.

It may be that improvement of 10% can also be achieved with the base load energy consumption for HWS in which case the intercept at the y axis would be at an average of 90 GJ/month instead of 100 GJ/month. The prevailing performance line can then be added as the current year proceeds and compared with the target performance line.

The drawback here is that it is not until well into the prevailing year that any discrepency is identified although the prevailing thermal performance line can be projected after the first two observations are entered and then adjusted as the prevailing year proceeds and more observations are available.

10.6   Continuous performance monitoring using mean daily outdoor temperature

The correlation between fuel or energy consumption and Degree Days has been tested from the historical observations given in Case study 10.1 and a close association confirmed for these two variables.

It can also be shown that a close association should exist between energy consumption and the 24-hour mean value of outdoor temperature preferably taken locally to the site. Mean outdoor air temperature can be measured daily from a maximum/minimum mercury in glass thermometer or its electronic counterpart in which case it can be automatically recorded on one of the output channels on a data logger and the weekly or monthly mean outdoor air temperature evaluated.

Alternatively mean outdoor air temperature can be determined approximately from the weekly or monthly Degree Day totals for the locality from:

$$t_m = t_b - DD/N \tag{10.5}$$

where t_m = average 24-hour mean outdoor temperature °C, t_b = Base temperature (taken as 15.5 °C for SDD) °C, N = number of days in period under review and DD = number of Degree Days to base t_b in period under review.

Monthly SDD totals are currently freely available from DEFRA's *Energy and Environmental Management* magazine.

The following case study analyses the thermal performance of a space heating plant with a HWS base load using average 24 hour mean daily outdoor temperature determined from weekly Degree Days. It will be shown that by adopting outdoor temperature a more detailed analysis of system performance can be made.

Case study 10.2

Consider for analysis the tabulated observations in Table 10.6 of SDD/week and energy consumption in MJ/week which includes a base load of 100 MJ/week attributed to HWS.

Solution

The energy consumption here is in MJ that can easily be converted to kWh since $1\,MJ = 0.278\,kWh$.

Adopting Equation 10.5 for the average 24 hour mean daily temperature t_m:

$$t_m = t_b - DD/N$$

The mean daily temperature average for each week can be determined and is shown in Table 10.7 together with the associated weekly energy consumption observations. Taking the independent variable x as the average 24-hour mean daily temperature and the dependent variable y as the weekly energy consumption in MJ, Table 10.8 can now be generated.

Table 10.6 Observations for analysis, Case study 10.2

Week no.	1	2	3	4	5	6	7	8	9	10	11	12
DD/week	79	91	31.5	70	21	61	3.5	55.3	13.3	47	24.5	17.5
MJ/week	475	525	290	360	150	312	195	245	110	335	300	240

Table 10.7 SDD/week converted to average 24-hour mean outdoor temperature from Table 10.6, Case study 10.2

Week no.	1	2	3	4	5	6	7	8	9	10	11	12
MJ/week	475	525	290	360	150	312	195	245	110	335	300	240
t_m		4.2	2.5	11	5.5	12.5	6.8	15	7.6	13.6	8.8	12	13

Table 10.8 Tabulated data based on observations in Table 10.7

x	y	xy	x^2	y^2
4.2	475	1995	17.64	225 625
2.5	525	1313	6.25	275 625
11	290	3190	121	84 100
5.5	360	1980	30.25	129 600
12.5	150	1875	156.25	22 500
6.8	312	2122	46.24	97 344
15	195	2925	225	38 025
7.6	245	1862	57.76	60 025
13.6	110	1496	185	12 100
8.8	335	2948	77.44	112 225
12	330	3960	144	108 900
13	240	3120	169	57 600
$\Sigma x = 112.5$	$\Sigma y = 3567$	$\Sigma xy = 28\,785$	$\Sigma x^2 = 1236$	$\Sigma y^2 = 1\,223\,669$

Adopting the simultaneous Equations 10.2 and 10.3 for linear regression and substituting for the 12 observations:

$$3567 = 112.5a + 12b$$
$$28\,785 = 1236a + 112.5b$$

Multiplying the first equation by 9.375

$$33\,441 = 1055a + 112.5b$$

Subtracting the second equation from the above we have

$$4656 = -181a$$

from which $a = -25.7$
Substituting for a in the first equation

$$3567 = -2891 + 12b$$

from which $b = 538$
The regression Equation 10.1 is therefore

$$y = -25.7x + 538$$

If x is now given two values, say 5 and 15, y is calculated from this regression equation as 409.5 and 152.5 respectively. These two sets of values can now be plotted on a graph of the average 24-hour mean daily temperature x against weekly energy consumption y and the best fit thermal performance line drawn as shown in Figure 10.6.

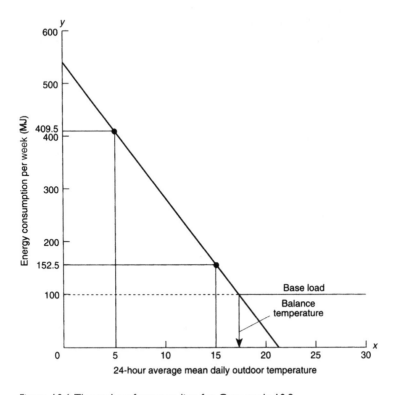

Figure 10.6 Thermal performance line for Case study 10.2.

You may also now like to plot the 12 observations on the graph to see how the performance line fits into the scatter of points.

Summary of Case study 10.2
From Figure 10.6 the results of the system performance analysis are:

Slope of performance line $= -25.7\,\mathrm{MJ/K}$

Temperature intercept can be measured from Figure 10.6 or calculated as follows: From the regression equation

$$y = -25x + 538$$

Thus

$$0 = -25x + 538$$

From which the temperature intercept, $x = (538/25) = 21.52\,°C$

The energy intercept by reference to Figure 10.6 or the regression equation is 538 MJ/week. The intercept is at zero Celcius.

If mean outdoor temperature for a week is $-2\,°C$, the energy consumption can be estimated from the regression equation where $y = -25 \times (-2) + 538 = 588\,MJ$. Or $588/7 = 84\,MJ/day$.

Balance temperature can be measured from Figure 10.6 or calculated as follows:

> From the regression equation $y = -25x + 538$
> Discounting the base load of 100 MJ and substituting: $100 = -25x + 538$
> From which $x = (438/25) = 17.52\,°C$.

- If the slope of the performance line becomes steeper this signifies a poorer system performance. If the thermal performance line becomes flatter this signifies an improvement in system performance. See Figure 10.6.
- The energy intercept indicates the energy consumption when the average 24 hour mean daily temperature is calculated as zero Celcius.
- The balance temperature is the outdoor temperature at which no heating is required; it is also the Base temperature t_b for the building where

$$t_b = t_i - d$$

and d = temperature rise due to indoor heat gains.

Thus Base temperature also is the outdoor temperature above which no heating is required. See Chapters 2 and 3.

If the balance temperature identified from the intersection of the thermal performance line with the base load is in excess of the Base temperature for the building, the calculated indoor heat gains may be in error or the space heating system is not functioning efficiently. The latter will certainly be the case if the balance temperature is greater than the design indoor temperature. Some reasons for this may be due to:

- Poor system performance resulting from inadequate space temperature controls, incorrect adjustment of space heating controls, malfunctioning of the space heating controls.
- Low boiler efficiency due to poor combustion conditions.
- Poor control of boiler plant causing unnecessary boiler on/off cycling.

The correlation coefficient can be determined from the summations in Table 10.8. It is not necessary here to extrapolate the base load from the weekly energy figures as it was in Case study 10.1. The reason for this is that the 12-week period over which the observations were taken did not include summer time when the space heating is shut down. In Case study 10.1 a calendar year's observations were taken which included the summer shutdown period for the space heating system.

Therefore from the correlation coefficient Equation 10.4 and the summations in Table 10.8

$$r = 28\,785 - 12(112.5/12)(3567/12)/$$
$$\sqrt{(1236 - 12(112.5/12)^2)(1\,223\,669 - 12(3567/12)^2)}$$
$$= -4656/\sqrt{(181.3 \times 163\,378)}$$
$$= -0.855.$$

The negative sign implies a negative association; see Figures 10.4 and 10.6.

From Table 10.1 the minimum correlation coefficient for there to be an association between the related variables of the average 24 hour mean daily outdoor temperature and weekly energy consumption is about plus or minus 0.8.

If the coefficient is calculated to be outside the minimum value the observations on which it is based are inaccurate and therefore the thermal performance line is invalid. More care would be needed in recording the observations and checks made on the measuring and recording equipment.

10.7   Correcting fuel/energy consumption to a common time base

Data from observations of energy consumption and Degree Days or mean daily outdoor temperature should always give a consistent indication of energy use. Sometimes occupancy and hence hours of use vary from one period to the next. It is important to check, for example, that an office space heating plant operates for regular periods each day of the week. Optimum start/stop time control of plant will have little effect except perhaps on Mondays after a weekend shutdown.

Consider observations of fuel consumption and mean daily temperature taken for a sports hall in Case study 10.3

Case study 10.3
The school sports hall has a heating system independent of the rest of the campus. It is heated by high temperature gas-fired radiant tube and the following observations were recorded (Table 10.9).

Table 10.9 Observations for Case study 10.3

Week no.	1	2	3	4	5	6	7
Period (hours)	35	30	30	40	30	45	30
Energy (GJ)	23.3	25	30	37.3	27	33	15
t_m (°C)	8	8.5	6	7.4	6.6	9.2	10

Solution

As the period of 30 hours is more regular the remaining observations can be corrected to the base of 30 hours

Week 1: $(23.3\,GJ/35\,h) \times 30 = 20\,GJ$
Week 4: $(37.3\,GJ/40\,h) \times 30 = 28\,GJ$
Week 6: $(33\,GJ/45\,h) \times 30 = 22\,GJ$

The revised observation record is shown in Table 10.10.

You may be able to deduce by inspecting the observations that there is a close association between the dependent and independent variables in Table 10.10. The scatter on a plot of mean daily temperature against weekly energy consumption should not be wide and therefore the line of best fit could be located manually without the need to undertake a linear regression analysis.

Plotting the observations on graph paper to suitable scales will prove the matter. This is done in Figure 10.7. The thermal performance line intersects the average 24 hour mean daily outdoor temperature axis and zero fuel consumption at approximately 16 °C. This temperature should be the balance temperature for the sports hall at which no heating is required. The thermal energy intercept when the average 24 hour mean daily outdoor temperature is zero is 47 GJ.

On the other hand you may prefer to undertake the regression analysis and the determination of the correlation coefficient which would validate the observations. Clearly if you have this on a database or spreadsheet there is no argument in favour of plotting the line by trial and error.

The linear regression equation from a solution by analysis is:

$$y = -3.25x + 50$$

Table 10.10 Corrected observations to a common time period

Week no.	1	2	3	4	5	6	7
Energy (GJ)	20	25	30	28	27	22	15
t_m	8	8.5	6	7.4	6.6	9.2	10

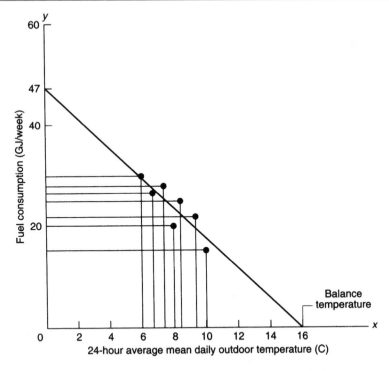

Figure 10.7 The thermal performance line for Case study 10.3 plotted by trial and
error.

If the independent variable x is now given the values of, say, 2 and 14°C,
the values of the dependent variable y will be 43.5 and 4.5 GJ respectively.
These values are then plotted on a graph – the average 24 hour mean daily
outdoor temperature on x axis and the fuel consumption in GJ on y axis –
as shown in Figure 10.8 from which the thermal performance line can now
be drawn and following data is obtained:

> The slope of the thermal performance line $= -3.25$ GJ/K.
> The temperature intercept (balance temperature) from the regression
> equation is $0 = -3.5x + 50$; from which $x = 50/3.5 = 14.3$°C (15.5 °C).
> The thermal energy intercept at a mean daily outdoor temperature of
> zero Celcius from the regression equation $= 50$ GJ
> The correlation coefficient is calculated as -0.916

Summary of Case study 10.3
You should now undertake the determination of the regression equation
and the correlation coefficient to confirm the solutions offered here. Clearly
more information is obtained by undertaking a full analysis. The balance

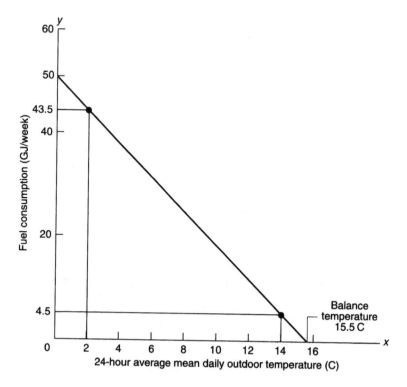

Figure 10.8 The thermal performance line for Case study 10.3 plotted from the linear regression equation.

temperature is more accurately determined along with the slope of the thermal performance line. The energy use on a day of severe weather (0°C) is predicted and the correlation coefficient is obtained. The minimum value of the coefficient for 10 observations from Table 10.1 is plus or minus 0.767. Seven observations were obtained in the study giving a coefficient of −0.916 which shows a close association between the two variables.

If the minimum value for the correlation coefficient for the number of observations as shown in Table 10.1 is not achieved the energy survey is void and greater care over measurement and collection of data is required.

10.8   Performance monitoring using cumulative sum deviation

It is sometimes important to be able to demonstrate to the client that an energy saving measure for space heating is actually saving energy following its provision. This is particularly true on a site that consumes energy for other purposes in addition to space heating.

Cumulative sum deviation (CUSUM) achieves this objective and can even be used to identify the monthly and cumulative savings in the annual fuel account. If CUSUM is adopted on a continuous basis it will show whether or not financial benefit resulting from the energy saving provision continues. CUSUM can also be adopted where energy is used for manufacturing products and in manufacturing processes.

Case study 10.4 which follows investigates the effects of introducing an energy saving measure on a space heating system by adopting CUSUM performance monitoring.

Case study 10.4
The first two columns in Table 10.11 list the monthly Degree Days x and energy consumptions y in GJ in the year before the energy saving provision is put in place for an office located in the Thames Valley. The table is then extended in order to undertake a linear regression analysis for that year.

Solution
Adopting the regression Equations 10.2 and 10.3 and substituting the summations for 12 observations where $n = 12$

$$230 = 1642a + 12b$$

and

$$44\,934 = 361\,108a + 1642b$$

Table 10.11 Monthly Degree Day, energy and regression data for Case study 10.4 one year before energy saving provision

Month	x	y	xy	x^2	y^2
September	51	10.0	510	2601	100
October	130	21.5	2795	16900	462
November	137	18.5	2535	18769	342
December	257	24.0	6168	66049	576
January	309	39.5	12206	95481	1560
February	223	29.0	6467	49729	841
March	277	32.0	8864	76729	1024
April	157	23.0	3611	24649	529
May	101	17.6	1778	10201	310
June	–	5.0	–	–	25
July	–	5.0	–	–	25
August	–	5.0	–	–	25
	$\Sigma x = 1642$	$\Sigma y = 230$	$\Sigma xy = 44\,934$	$\Sigma x^2 = 361\,108$	$\Sigma y^2 = 5820$

Multiplying the first equation by 136.8 we have

$$31\,464 = 224\,626a + 1642b$$

subtracting this from the second equation we have

$$13\,472 = 136\,482a$$

from which $a = 0.1$
Substituting for regression coefficient a in the first equation we have

$$230 = 162 + 12b$$

from which $b = 5.66$
Thus the regression equation is $y = 0.1x + 5.66$.

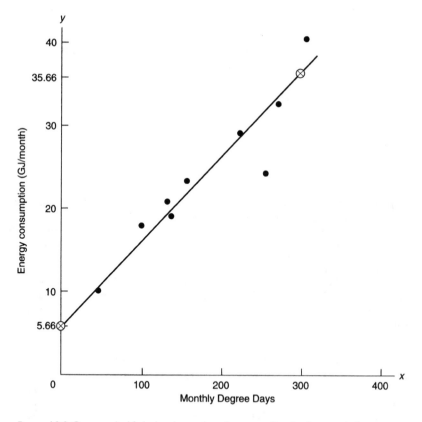

Figure 10.9 Case study 10.4: the thermal performance line in the year before energy
saving provision

Table 10.12 Summations of weather related data Case study 10.4

Months	x	y	xy	x^2	y^2
September to May	$\Sigma x = 1642$	$\Sigma y = 215$	$\Sigma xy = 44\,934$	$\Sigma x^2 = 361\,108$	$\Sigma y^2 = 5745$

The performance line of best fit can now be drawn for the year before the energy saving provision although it plays no part in the calculation of CUSUM. This is shown in Figure 10.9 taking two values: when $x = 0$, $y = 5.66$ and when $x = 300$, $y = 35.66$.

Before proceeding further it is advisable to calculate the correlation coefficient from the weather related data in Table 10.11 to ensure that the observations are sufficiently accurate. There are nine observations of weather related data and the summations are listed in Table 10.12.

You will notice in Table 10.12 that only summations for y and y^2 have altered from Table 10.11 on account of the summer months when the energy consumptions are not weather related.

The correlation coefficient r for the weather related observations in the first year can now be determined from Equation 10.4:

$$r = (44\,934 - 9(1642/9)(215/9))/$$
$$\sqrt{(361\,108 - 9(1642/9)^2)(5745 - 9(215/9)^2)}$$
$$= (44\,934 - 39\,226)/\sqrt{(361\,108 - 299\,559)(5745 - 5136)}$$
$$= 5708/\sqrt{(61\,549 \times 609)}$$
$$= 5708/6122$$

Thus the correlation coefficient $r = +0.93$.

The minimum value for the coefficient for nine observations is, from Table 10.1, about plus or minus 0.8. Thus the observations are validated and we can proceed.

The calculation of CUSUM for the office can now be undertaken. This is listed in Table 10.13 which includes the monthly Degree Days and energy consumptions for both the year before and the year following the intoduction of the energy saving provision.

The column of predicted energy consumption is calculated using the regression equation determined from the observations made in the year prior to the introduction of the energy saving measure. The equation is $y = 0.1x + 5.66$.

You should now confirm agreement with the regression equation.

The column headed "difference" identifies the difference between actual energy consumption y and predicted energy consumption. The minus sign

Table 10.13 CUSUM for Case study 10.4

Month	DD x	Actual energy y	Predicted energy	Difference	CUSUM
September	51	10	10.76	−0.76	−0.76
October	130	21.5	18.66	2.84	2.08
November	137	18.5	19.36	−0.86	1.22
December	257	24	31.36	−7.36	−6.14
January	309	39.5	36.56	2.94	−3.20
February	223	29	27.96	1.04	−2.16
March	277	32	33.36	−1.36	−3.52
April	157	23	21.36	1.64	−1.88
May	101	17.6	15.76	1.84	−0.04
June	–	5	5.66	−0.66	−0.70
July	–	5	5.66	−0.66	−1.36
August	–	5	5.66	−0.66	−2.02
September	55	8.7	11.16	−2.46	−4.48
October	138	13	19.46	−6.46	−10.94
November	241	17	29.76	−12.76	−23.7
December	299	22	35.56	−13.56	−37.26
January	337	28	39.36	−11.36	−48.62
February	308	28.3	36.46	−8.16	−56.78
March	270	21.5	32.66	−11.16	−67.94
April	199	18	25.56	−7.56	−75.50
May	114	11.2	17.06	−5.86	−81.36
June	–	5	5.66	−0.66	−82.02
July	–	5	5.66	−0.66	−82.68
August	–	5	5.66	−0.66	−83.34

indicates that the predicted energy consumption exceeds the actual monthly consumption of energy. The last column is the cumulative effect (CUSUM) of the "difference" column over the two consecutive years. The CUSUM values in the last column of Table 10.13 are now plotted against the months of the 2-year period and the result is shown in Figure 10.10.

Analysing the plot of CUSUM in Figure 10.10, Case study 10.4
- The graph is based upon the regression equation calculated from the observations prior to the energy saving provision.
- You will see that the plot hovers around the base line during the year preceeding the implementation of the energy saving provision. This is to be expected since the regression equation used for all the predictions was calculated from that year's observations.
- The deviation begins in the September of the second year and continues until the following May at the end of the heating season. The deviation reverts at this point back to a second base line. This occurs because the office base load which accounts for the consumption of hot water is weather unrelated and therefore not influenced by Degree Days.

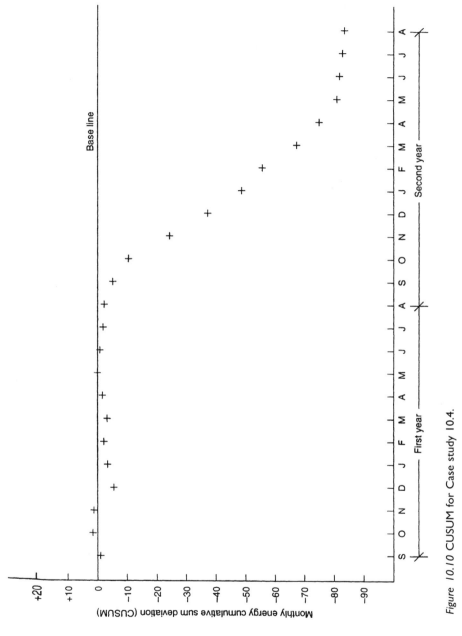

Figure 10.10 CUSUM for Case study 10.4.

- The deviation which occurs between the September of the second year and the following May identifies the cumulative savings in the monthly energy consumption, the total being in the region of 83 GJ for that period. This can easily be costed progressively each month starting in the September of the second year.
- If the CUSUM calculation is continued into the year after and beyond, a constant watch can be taken to ensure that the savings continue. If the slope of the line of deviation in Figure 10.10 changes during the following heating season it shows that either further energy savings have been introduced (steeper slope) or that the initial energy saving provision is lapsing (flatter slope) or that something else has caused an increase in energy consumption.
- CUSUM cannot divulge what the energy savings or increased energy consumption can be attributed to. In fact in this case study which relates to an office block in the Thames Valley, temperature controls were upgraded and commissioned in the August at the end of the first year of observations. As long as the temperature controls are maintained in accordance with operating procedures and preventive maintenance to the building and to the services is ongoing the savings in energy should continue. This will be confirmed by the CUSUM deviation slope.

10.9   Chapter closure

You now have the underpinning knowledge to undertake a variety of monitoring and targeting procedures. You can analyse the results of monitoring energy consumption and prepare documentation to support evidence of the effects of energy conservation measures, and documentation to confirm savings in the cost of energy consumed for space heating and HWS.

Chapter 11

Regulations relating to greenhouse gas emissions

Nomenclature

CCL   Climate Change Levy
ETS   Emission Trading Scheme
EU ETS   European Emission Trading Scheme
EMS   Environmental Management Systems
ETL   Energy Technology List
ECAs   Enhanced Capital Allowances
C&C   Global Contraction and Convergence
NAP   National Allocation Plan
SMEs   Small and Medium Sized Enterprises
EPBD   European Energy Performance in Buildings Directive
ROCs   Renewable Obligation Certificates
OFRs   Operating and Financial Reviews
SAP   Standard Assessment Procedure
BETTA   British Electricity Trading and Transmission Arrangements
LEC   Levy Exemption certificates

11.1 Introduction

The Kyoto Protocol regulates emissions of six greenhouse gases including carbon dioxide, which is the most important one, as well as methane, nitrous oxide, hydrofluorocarbons, perfluorocarbons and sulphur hexafluoride.

The Government's Climate Change Agency (CCA) is integrating climate change protection measures across activities aiming to reduce emissions by 20% by 2010 relative to 1990 and 60% by 2050 relative to 2000.

There are two issues that will be focused upon in this chapter. The various acts, directives and regulations relating to emissions of greenhouse gases originating from government and the European Union and the Building Regulations Directive Part L2, 2002 and 2006.

The issues surrounding greenhouse gas emissions are still evolving and will continue to evolve beyond the scope of this publication. At present reductions in greenhouse gas emissions relate to emissions of the gas carbon dioxide. Sulphur dioxide and Nitrous oxides will be the next emissions to be included in the reduction programme. However, it is hoped that much of the material included in this chapter will be relevant in an historical context and for some time to come.

The new Building Regulations are to be published in April 2006 following closely on the Building Regulations of April 2002. There are driving forces behind these two issues. The first is the fact that man is consuming the Earth's finite resources at an exponential rate that does not bode well for future generations. The second is the ever increasing evidence that the Earth's climate is changing and becoming more unstable partly, if not primarily, due to the effect of greenhouse gases and holes in the ozone layer.

11.2   The Climate Change Levy

The UK Government introduced the Levy in April 2001. It is essentially a tax on non-domestic fuel consumption and was intended to be fiscally neutral.

The reimbursements on National Insurance Contributions designed to offset tax liability will, however, ultimately prove beneficial to companies with high staff levels and low energy consumption as in the service industries. Table 11.1 gives the details of the Levy Rates at 1 April, 2001, which will increase or decrease with time.

Example 11.1
Consider Case study 3.2 that concerns the refurbishment of an office block in Manchester to a naturally ventilated open plan Type 2 office. Given the cost of natural gas as 3.4 p/kWh and electricity as 10 p/kWh, determine the Levy and compare with the annual cost of the energy consumed.

Solution
From Case study 3.2,

Table 11.1 The impact of the Climate Change Levy

Fuel	Levy rate
Gas	0.15 p/kWh
Coal	1.17 p/kg
Electricity	0.43 p/kWh
LPG	0.96 p/kg

National Insurance Reduction
Employers' contributions: 0.3% from 2000/2001 level

For space heating AEC = 171 116 kWh
For hot water supply AEC = 8059 kWh
Total AEC = 179 175 kWh

Annual fuel consumption AFc = $179\,175 \times (3.4/100) = £6092$
From Table 11.1 the annual Levy for gas = $179\,175 \times (0.15/100) = £269$
Taking the electricity consumption as "Good Practice" for a Type 2 office
of 54 kWh/m^2 AEC = $54 \times 6 \times 370 = 119\,880$ kWh
Annual electricity consumption AEC = $119\,880 \times (10/100) = £11\,988$
From Table 11.1 the annual Levy for electricity = $119\,880 \times (0.43/100) =$
£515
Total annual Levy = $£269 + £515 = £784$ or 35.3 p/m^2
Total annual cost of energy = $£6092 + £11\,988 = £18\,080$ or £8.14/m^2

Summary of Example 11.1
The annual cost of the Levy of £784 may be offset by the 0.3% reduction
in National Insurance Contributions. This will of course depend on the
number of employees occupying the building, which in this case study was
estimated at 222.

Example 11.2
From Appendix 4 the following benchmarks related to a naturally ventilated open plan office (a) and a prestige air-conditioned office (b) are listed
in Table 11.2. Using the figures in Table 11.1 calculate the annual Climate Change Levy in p/m^2 of floor for these two office types for gas and
electricity.

Solution
The solutions are: (a) 59.2 p/m^2, 35.07 p/m^2 and (b) 185 p/m^2,
117.72 p/m^2. You should now confirm agreement to these solutions.
   The Levy is paid by industry, commerce and the public sector that includes
local authorities, for non-domestic energy use.
   A number of manufacturing industries with high energy consumption
and relatively fewer employees have offset potential losses by investing in
new plant and advanced technology and as a consequence have achieved
significant reductions in fossil fuel consumption.

Table 11.2 Benchmark data from Appendix 4

Building	Fuel	Typical (kWh/m^2)	Good practice (kWh/m^2)
(a)	Fossil fuel	151	79
	Electricity	85	54
(b)	Fossil fuel	210	114
	Electricity	358	234

There are four main steps that can be taken by participating organisations to reduce the impact of the Levy.

- Electricity produced from renewable sources except hydro-electric over 10 MW will be exempt.
- Electricity generated on site for use by the owner is exempt.
- Heat generated by and the fuel input for "good quality" CHP will be exempt as will the electricity generated from CHP. To qualify for exemption CHP schemes must be registered as "good quality". Self-assessment forms are available from the Combined Heat and Power Quality Assurance Programme.
- Where an organisation finds the income from National Insurance reductions outweigh the cost of the Levy, this income can be used to promote energy efficiency and consequently reduce energy consumption.

11.3   The Energy Technology List

To coincide with the launch of the CCL a list of energy saving products qualifying for Enhanced Capital Allowances (ECAs) administered by the Carbon Trust was announced. The Energy Technology List (ETL) initially contained 1200 individual products in eight technologies:

- Combined heat and power
- Refrigeration
- Motors
- Boilers and add-ons
- Pipe insulation
- Lighting
- Variable speed drives
- Thermal screens.

The scheme aims to support investment in these low carbon technologies and attracts ECAs. Since its introduction further products have been added to the ETL.

11.4   Enhanced Capital Allowances

For a profit making company ECAs provide a tax break enabling 100% of the cost of investing in new plant and equipment, that must be included in the Energy Technology List, to be written off against tax in the first year of purchase.

The Carbon Trust has been appointed as the administrators of ECAs and the scheme for energy saving investment applies to accounting periods ending on or after 1 April 2001 (corporation tax) and periods ending on or after 6 April 2001 (income tax).

Product manufacturers have shown significant interest in ECAs by applying to be included in the ETL.

11.5 The Carbon Trust

The Carbon Trust was set up by the government in July 2000. It was part of the CCL package to join up a programme including research and development, fiscal incentives and advice to encourage businesses, especially small and medium sized enterprises (SMEs) to implement energy efficiency measures.

The Carbon Trust serves as the focus for strategic and executive action to ensure business adapts successfully to the challenges presented by climate change. The Trust is a non-profit company limited by guarantee and is one of the consequences of international progress on action to tackle climate change, by reducing carbon use in energy, following a meeting at the Hague in November 1999.

11.6 Carbon or carbon dioxide?

Greenhouse gas emissions at present relate to emissions from the gas carbon dioxide. There are two terms in use here in defining emissions, namely carbon and carbon dioxide and it is necessary to know the relationship of one with the other. Emissions of carbon dioxide can only derive from the complete combustion of carbon. Natural gas for example has typical constituents of methane, ethane, propane, butane, nitrogen and carbon dioxide, in varying proportions. The first four constituents contain carbon and hydrogen. The constituents nitrogen and carbon dioxide pass through the combustion process unaffected. The carbon content of each constituent follows the chemical equation by mass:

$$C + O_2 \rightarrow CO_2$$

Using the molecular mass of carbon, oxygen and carbon dioxide we have

$$12 + 32 \rightarrow 44$$

So 12 kg of carbon requires 32 kg of oxygen for complete combustion to produce 44 kg of CO_2.

Thus the important ratio here to convert carbon to carbon dioxide is 44/12. Alternatively to convert carbon dioxide emissions to carbon, the ratio is 12/44. Both terms, carbon emissions and greenhouse gas (carbon dioxide) emissions, are in use and need to be understood. For example, an annual emission of 10 tonnes of carbon is equivalent to 36.67 tonnes of carbon dioxide using the ratio 44/12.

Combustion of four fossil fuels is presented in detail in another publication in the series.

11.7 Emissions Trading Scheme

The idea of emissions trading is to achieve cuts in emissions at lowest cost to the economy. Each installation that comes under the scheme will be allowed to emit a certain amount of carbon dioxide. The allocations are free and the quantities are based on historic emissions – a technique known as "grandfathering". If plants overshoot, they will have to acquire further emissions allowances, probably by buying them on the open market. If they use less than their total number of allowances, they can sell the remaining to others. In theory this should be the economically most efficient way of achieving savings because companies will reduce emissions through improved efficiency or investment in new technology if this works out cheaper than buying the required number of extra allowances. If purchasing allowances is cheaper they will do this instead. Either way they meet their emission allowance targets at the lowest net cost. Those who invest to bring their emissions below their target can sell the unwanted allowances to offset those investments.

There are financial penalties for failure to balance the books: 40 Euros per tonne in the first phase of the scheme 2005–2007 and 100 Euros per tonne between 2008 and 2012. Operating without a permit, contravening the terms of that permit and misleading or misreporting emissions data are all criminal offences under the law. In the UK, failure to correct data that is found incorrect will also be a criminal offence.

The publication of the UK National Allocation Plan (NAP) for the European Emissions Trading Scheme (EU ETS) was launched in April 2004. The first phase of EU ETS runs for 3 years from 2005 to 2007. Thereafter trading periods will last for 5 calendar years.

It is interesting to note that Europe's 160 million buildings are responsible for 40% of total energy consumption.

Every member state of the EU has drawn up a similar NAP and submitted it to the European Commission. The NAP sets out the total number of emission allowances, each representing 1tonne of CO_2, to be allocated to the industry sectors covered by the EU ETS.

The scheme will impose requirements on the largest emitters of carbon dioxide to monitor and account for their emissions. The industries covered are:

- Combustion installations
- Mineral oil refineries

- Coke ovens
- Metal ore processing
- Iron and steel production
- Cement and lime production
- Glass manufacture
- Ceramics manufacture
- Pulp, paper and board production.

Together, these account for about 50% of all UK carbon dioxide emissions. On 1 January 2005, operators of installations in these sectors were given a greenhouse gas emission permit and by 30 April 2006 these permit holders will have to declare emissions levels to the EU. Clearly they will have to monitor emissions of specified gases (only CO_2 for the first phase), have emissions verified independently by an accredited body and submit a report on this data.

The UK launched the EMS in September 2004 to its industrial sectors. If after 3 years emissions have increased, further permits will have to be bought. If emissions have decreased the surplus permits can be sold to those more profligate industries thus initiating EMS. Refer to Appendix 10.

The EU ETS is therefore a cap and trade scheme in which a company is free to trade with any other company in the EU. This means that a permit holder can either decide to reduce its emissions levels, buy allowances through the open market to bridge the gap between target and actual emission levels or use a combination of both methods.

Operators that achieve better emission reductions than their target can sell the savings on the open market.

11.8   Levy Exemption Certificates

Renewable power generators are required to apply for Levy Exemption Certificates (LEC) to prove that the energy comes from renewable sources. These Certificates are then sold along with the green electricity to the electricity supplier. The supplier uses the LECs to demonstrate to Customs and Excise that its electricity has been generated from qualifying sources and is eligible to be sold to customers without having to pay the Climate Change Levy.

11.9   Renewable Obligation Certificates

The Renewables Obligation Order was introduced by the government in 2002 for increasing the contribution of energy from renewable sources to fulfil the EU Renewables Directive.

The RO requires licensed electricity suppliers to source a percentage of their sales from eligible renewable sources. This percentage starting at 3% increases each year and reaches 10.4% by 2010.

Each MWh of renewable electricity generated is accompanied by a Renewables Obligation Certificate (ROC). At the end of each compliance period, which is 1 year, electricity suppliers must provide the electricity regulator with the appropriate number of ROCs to prove that they have fulfilled their RO. If they cannot produce the required number of ROCs they pay a buy-out fee of £30 for every MWh that they are short.

ROCs in this unique system are the currency of the Renewables Obligation Order. They can be traded with suppliers buying them from generators, brokers or other suppliers. The floor price is set by the buy-out fee described and is index linked. The ceiling price depends upon the excess of demand for ROCs over supply.

When insufficient electricity is generated from renewable sources to meet the demand from suppliers as was the case up to 2003 then the value of the ROC goes above the floor price originally set at £30.

This is because the buy-out fund into which suppliers of electricity pay if they fail to meet their targets is redistributed back to suppliers in proportion to their level of compliance under the RO. This recycling value is added to the floor price giving a higher overall ROC value. The average price at auction in February 2003 was £65/MWh.

The recycling value per ROC redeemed is worked out as follows:

Recycling value per ROC redeemed = (total buy-out fund)/

(total number of ROCs redeemed by all suppliers)

11.9.1 Evaluating ROCs

Two examples show how this works:

- If only half of the required renewable electricity is available then ROCs will be worth:
  The base value of £30 + £30 received in recycled buy-out money by the supplier for each ROC redeemed, namely £60.
- By the same token if there is a 25% shortfall then ROCs are worth:
  £30 + £15 of recycling benefit, namely £45.

These certificates may be of use to SMEs, the private housing sector, housing associations and local authority housing where small scale renewable generation of electricity and domestic hot water is installed on site. An example might be the use of the small plug in wind generator.

11.9.2 Purchasing green electricity

As identified in Section 11.8 customers can purchase green electricity that may be generated from wind turbines, hydroelectric, biomass, landfill gas, industrial and municipal waste and sewage gas. However, some of these renewable sources are not CO_2 neutral. The purchaser may want to establish with the supplier what the mix of green electricity being offered includes and in what proportions. The purchaser will also want to know the surcharge for the green electricity so that it can be compared with the CCL otherwise payable at 0.43 p/kWh (see Section 11.2).

11.9.3 Base load generation

Electricity generation comes from an increasing mix which includes that from renewable sources such as wind turbines and hydroelectric. However, there is a requirement for the generation of a consistent national base load that is at present provided largely by nuclear power stations most of which are coming towards the end of their life cycles and will have to be decommissioned.

The generators therefore have to consider what can replace the current nuclear power plants to provide a continuous base load for the UK.

11.10 The EU Energy Performance in Buildings Directive

The EPBD has far reaching implications for energy managers, facilities managers, owners, operators and developers of all buildings in the UK. This of course must include, by implication, building services engineers.

The EPBD consists of 17 Articles 7 of which directly addresses building services engineers:

Article 1   List of energy requirements
Article 3   Energy performance of systems
Article 5   Energy systems for new buildings
Article 7   Performance certificates
Article 8   Boilers
Article 9   Air-conditioning
Article 10   Independent certification

The EU Directive adds underlying support to the UK government's Energy White Paper published in 2003.

To satisfy the terms of the Directive, legislation must be in place by 4 January 2006.

Key provisions of the Directive are:

- Minimum requirements for energy performance of all new buildings.
- Minimum requirements for the energy performance of large existing buildings subject to renovation.
- Energy certification of all buildings with public buildings being required to display the energy certificate.
- Regular and mandatory inspection of boilers and air-conditioning systems in buildings.

To help the government to implement the Directive, a working group has been established that includes representation from CIBSE and the Energy Institute. The Energy White Paper commits government to transpose the EPBD into law by the end of 2005.

11.10.1   New buildings

For new buildings over 1000 m² floor area the Directive requires formal consideration be given at the design stage to alternative systems of heating and cooling based upon decentralised renewable energy supply systems.

11.10.2   Refurbished buildings

For buildings being renovated and over 1000 m² floor area the energy performance is upgraded to incorporate all cost-effective energy efficiency measures.

11.10.3   Building energy certificates

The EPBD requires that whenever a building is constructed, sold or rented out, a certificate detailing its energy performance must be made available to the owner or by the owner to the purchaser or tenant. It must include:

- Current benchmarks
- Recommendations for cost-effective energy performance improvements.

Initially the Energy Certificate will be required for sale or let of nearly all buildings and displayed prominently in larger public buildings. A sample Energy Certificate is given in Figure 11.1.

11.10.4   Building log books

Log books for commercial and industrial buildings, although not a specific requirement under the EPBD, are a requirement of Part L2 of the

Building energy Performance	As built	In use
Very energy efficient		
A		
B	B	
C		
D		D
E		
F		
G		
Not energy efficient		
Asset Rating Method: UK		
National Standard 2004	Calculated	Actual
Units used kg CO_2/m^2 annum	48	83
Occupancy level m^2/person	14	12
Equipment heat gain W/m^2	12	12
Weekly occupancy hours	55	58
Heating performance ratings	B	C
HVAC performance ratings	C	C
Lighting performance ratings	A	B
Management rating (in use only)		F
Internal environment quality	not assessed	
Risk level	not assessed	
GB 2004		
Certification organisation:	address, contact and email	

Figure 11.1 Sample non-domestic building energy certificate.

UK Building Regulations. They apply to new and refurbished buildings in England and Wales. They are also required when plant such as boilers and air handling units are upgraded or replaced in existing buildings.

Log Book templates are available for facility managers from CIBSE along with a tool kit that includes a log book template and three completed example log books.

11.10.5 Plant inspection

The EPBD provides member states with two options for heating systems:

- To establish a regular inspection of boilers.
- For boilers larger than 20 kW and over 15 years old there must be an inspection of the entire heating system.
  Option two includes:
- Expert advice to users on modifications, replacement and alternative solutions with governments producing a report every 2 years showing how this achieves as much as the first option.
  For air conditioning systems governments must:
- Establish regular inspections of plant with an effective rated output of more than 12 kW.
- Assessment of the efficiency and sizing of the air-conditioning compared with the cooling requirement of the building.
- Appropriate advice must be provided.

11.10.6 Independent experts

EU Member States must ensure that certification of buildings, the drafting of recommendations and inspection are carried out in an independent manner by qualified and accredited experts.

11.10.7 Conclusions

It can be seen from the earlier sections in this chapter that much of the preparation to implement the EPBD has already been, or is being, put in place by the UK Government and associated organisations.

11.11   Domestic dwellings

11.11.1 New dwellings

The procedures given in the EU. EPBD is reflected in the latest Building Regulations. New dwellings are subjected to Part L1 of the Building Regulations 2005. The performance standard (Standard Assessment Procedure or SAP) is based upon a target carbon dioxide emissions rating for space heating, hot water and lighting. This allows developers flexibility in the way the target is achieved. For example an emphasis on thermal insulation of the dwelling envelope and natural lighting on the one hand or concentration on the heating, hot water and lighting systems using high efficiency plant and passive or combined systems for heating and power (CHP) on the other.

Another feature of the Regulations is the need for pressure testing the new dwelling to confirm the appropriate standard of air tightness has been achieved. Air leakage (infiltration and exfiltration) can be a significant source of energy loss. Clearly air tightness must be coupled with effective ventilation to ensure adequate indoor air quality. Ventilation for health is covered in Part F of the Building Regulations.

11.11.2 Existing dwellings

The headline feature in the Regulations is that more work in existing buildings will be subject to the Building Regulations. By capturing this additional work when perhaps only part of an existing building is being renovated it is possible to incorporate energy efficiency improvements at the most cost effective point. In the case of private ownership this may be at the point when the dwelling is made available for sale or let. The Energy Performance Certificate for the dwelling will be provided to the intending purchaser or tenant and will include a list of cost effective improvements that could be made.

It is not the case that existing dwellings must ultimately be renovated to the standard of new dwellings but that improvements shall be reasonable given the age of the building and that they are cost effective in terms of energy reduction.

11.12 Contraction and convergence

The UK based Global Commons Institute (GCI) first proposed C&C in 1990 and has since refined the concept. This is essentially linking the reduction in global greenhouse gas emissions with global security and global equity.

It calls for a reduction in carbon dioxide emissions of 85% by the year 2050. This compares with the 60% reduction target of the Government.

11.13 British Electricity Trading and Transmission Arrangements

BETTA which formed part of the Government's Energy Bill that reached the statute book in July 2004 replaces the New Electricity Trading Arrangements (NETA). NETA has fundamentally changed the purchasing strategies for gas and electricity. Whereas in 2001 when NETA was introduced an energy supplier would hold a quoted price for 2–3 days, offers now are open for just 2–3 hours. This is the consequence of moving from policy driven to market driven purchasing.

The energy manager or facilities manager is now required to respond to rapidly changing market forces when purchasing gas or electricity from the supplier. This can be complicated if the purchasing organisation is pan-European as the energy manager will have to deal with different regulatory regimes and maybe different languages. The other effect of the move from a policy driven market to a futures market is that the Electricity Generators have difficulty in finding finance investment in spare generating capacity that has always been available in the past when the industry was nationalised.

11.14 Towards sustainable reporting

Operating Financial Reviews are meant to stimulate an organisation's awareness of sustainability issues.

Companies quoted on the Stock Exchange from financial years beginning on or after the 1 January 2005 are required to produce OFR's for shareholders. OFR's account for the social and environmental performance of the company. They are published separately from annual reports to shareholders so that they can be made available to the public at large on a freedom of information basis. A typical OFR may have the following headings:

11.14.1 Environmental

- Global warming: carbon dioxide emissions
- Waste: total generated, hazardous waste, recycled waste
- Pollution: CFC's consumption, HCFC's consumption
- Resources: oil consumption, gas consumption, coal consumption, water consumption
- Prosecutions: details of any environmental prosecutions
- Benchmarking: current and forecast benchmarks in kWh/m^2.

11.14.2 Social

- Health and safety: major accidents, fatalities at work, safety prosecutions
- People: total number of employees, employees under 18 years, staff turnover, ratio of male to female staff
- Work: number of days lost in industrial disputes, number of days lost due to sickness
- Training: number of training days, spend on training
- Prosecutions: number
- Charitable donations: total, gifts in kind.

Clearly the introduction of OFR's sends a strong message to quoted companies to take an active interest and ownership of their social and environmental responsibilities. Some enlightened organisations have been publishing annual environmental and social responsibility reports for a number of years.

11.15 Chapter closure

The reader will now have substantial background knowledge of the regulations and directives that apply to climate change. Some of these regulations and directives will affect the building services engineer in the processes of system selection and design, installation, commissioning, forecasting energy consumption, auditing and benchmarking. Not least is the need for qualified and accredited building systems inspectors.

Other regulations and directives given in this chapter affect the work of the energy manager or facilities manager who is tasked with monitoring and targeting building energy performance, preparing and costing energy efficiency programmes and operating the systems maintenance repair, renewal, recycling and waste disposal procedures.

Chapter 12

Trends in building services

Nomenclature

WLC   Whole life costs
PFI   Private finance initiative
PPP   Public private partnership
PPM   Planned preventative maintenance
PCM   Performance conditioning monitoring
VSD   Variable speed drives
EU   European Union
CHP   combined heat and power

12.1 Introduction

There is a clear message emanating from the Chartered Institution of Building Services Engineers and other professional bodies that the construction industry and hence the building services industry must account for the environmental impact that initial decisions have on building, refurbishing, servicing and operating buildings. We have to move away from adversarial procurement and the capital cost justifier to a more sophisticated philosophy that engages social, environmental and sustainable issues.

There are four issues relating to the more enlightened procurement process:

- The impact of sourcing the materials, products and land for the project
- The impact of building and servicing the project
- The impact of operating the project over its life
- The impact of refurbishing/decommissioning the project and the disposal of material.

It is estimated that at the end of the life of a building, its life cycle costs will be at least five times the initial capital cost. These two costs added together represent the whole life cost (WLC) for the building.

The cost of demolition, re-use, recycling and waste disposal must be included. Taking the sustainable view, material designated for waste disposal must be biodegradable otherwise it must be re-used or recycled or it will attract heavy penalties in land fill charges.

Some owners are now taking a keen interest in the concept of sustainable building and the costs over the life of the building and are prepared to accept higher capital costs if low operational costs can be designed into the building initially. This approach forms part of the concept of sustainable development. A complete change is required in the way "products" are sourced, in terms of raw material, and manufactured. Along with this is the need for society to completely rethink the way it consumes the necessities and luxuries of life that involves packaging, product life, re-use, recycling and energy consumption during the product life cycle.

The difficulty facing the industrialised nations is that we have got used to consuming the Earth's finite raw materials as though they were infinite. Indeed western economics is based upon consumption. Western society must consume to exist. This is compounded by the current throw away society. There was a time that ended about 40 years ago when many products at the end of their useful life could be traded in for new or reconditioned replacements. The phrase "part exchange" included many parts of products that could be returned and reconditioned. This principle will need revisiting in the light of modern technology.

12.2 Sustainable development

Sir Jonathon Porritt, chairman of the Sustainable Development Commission, gave an address to the Institution of Incorporated Engineers in 1999 related to an engineering-focussed view of sustainable development.

His seven basic principles for what he called eco-efficiency were timely and thought provoking:

- Reduce the material intensity of products.
- Reduce energy intensity.
- Reduce toxic dispersal.
- Enhance material recyclability.
- Maximise the sustainable use of renewable resources, such as energy, and raw materials, such as biomass.
- Extend the durability of the product.
- Increase maintainability so that the use of the product is increased for a lower resource throughput.

It might be argued that these principles lie at the door of product manufacturers. However, the mechanical and electrical services in the building should be collectively considered as products, since once installed, they require energy to operate and maintenance, both of which have a direct bearing on six if not seven of these principles. The difficulty facing the industrialised nations is that we have got used to extracting and processing raw materials at a greater rate than Nature can replenish them. Sir Jonathon believes that scientists and environmentalists are now beginning to converge around an understanding of what the basic scientific principles relating to the activities of humans on their world are.

- The conservation of energy and matter
- Energy and matter tend to disperse over time
- Energy and matter cannot disappear
- Increases in net material quality are almost exclusively derived through photosynthesis.

The reader will no doubt see the principles of thermodynamics and cell biology here:

- The first scientific principle means that waste material and the products of combustion from fossil fuels do not disappear and that the entire concept of waste disposal and the disposal of combustion products is an illusion. Approximately one hundred tonnes of raw material along with the energy from fossil fuels enters the industrial process to generate one tonne of product.
- The second scientific principle relates to the dispersal of energy and matter. Energy derived from fossil fuels produces products of combustion that disperse, with consequences, into the atmosphere. Natural resources that are mined and extracted eventually disperse back into nature. Steel eventually rusts, for example. Neither the products of combustion nor rust can return to fossil fuel or steel.

   There are some products that do not disperse or degrade of course. Glass and some plastics are examples.
- The third principle is that matter and energy cannot disappear. What society consumes is not the stuff itself but the products made from it. If society consumes our natural resources faster than the Earth can supply them it is obviously becoming poorer and the serious matter of dealing with the products of combustion and waste matter will not go away.
- The fourth principle is one of cell biology and asserts that increases in net material quality are almost exclusively derived through sun-driven processes – essentially through photosynthesis.

12.3 Whole life costing

Whole life costing (WLC) includes capital costs *and* life cycle costs and has developed from the procurement routes for public buildings such as schools and hospitals and involves two approaches, namely the private finance initiative (PFI) and the public private partnership (PPP).

Methodologies for planning and predicting the life of building services are identified by ISO 15686. Quality assurance for whole life cost models is also included together with how to test them to establish the uncertainties and risks inherent in them.

The effects of system design on energy consumption and cost are addressed by CIBSE TM 30.

Whole life costing is used to help determine the basis for an overall contract to design, build, operate and maintain the project over a contract period of 30–40 years. It is an exercise in risk management appraisal, a relatively new context in which risk management has an important role to play.

The Construction Research and Innovation Strategy Panel defines WLC as "the systematic consideration of all relevant costs and revenues associated with the acquisition and ownership of an asset". You will notice that this definition also includes the "revenues" obtained from the activities undertaken within the building during its life.

The elements of WLC include:

- Capital costs
- Life cycle costs
- Identifying best value
- Maintenance of M&E systems
- Supply chain management.

We discuss each of these below.

12.3.1 Capital costs

Design costs, installation costs, commissioning costs, handover costs.

12.3.2 Life cycle costs

Energy costs, operating costs, maintenance and repair costs, down time costs, environmental and social (environmental impact) costs, decommissioning, recycling and disposal costs.

Life cycle costing as applied to mechanical and electrical services within the building seeks to determine the best through life business model for installations using a wide range of techniques and disciplines. Clearly the ideal is to aim for low capital cost and high savings in use and final removal.

At its most benign the life cycle model will include the cost of dismantling, recycling and waste disposal along with the total environmental impact of procurement and use. This will include the impact of using raw materials and fossil fuel during product manufacture (known as "embodied energy") and burning fossil fuel during the life of the project.

To minimise WLC, life cycle costing normally focuses on the 1:5 ratio for capital cost to life cycle cost, for an office block with a 20-year life. From this perspective it can be seen that good design that takes into account operating costs will have a significant payback throughout the life of the project. Conversely poor design and lack of knowledge by operators on how the mechanical and electrical systems work will add to operating costs.

Traditional procurement has ignored the link between the decisions made during the design and install stages and their effect on operating costs.

Estimation of operating and final removal costs for a new project being undertaken in the traditional way is rarely sought.

12.3.3 Identifying best value

Taking the WLC model, there are four principles that can be applied to identify the building services components and systems having best value.

1 Maximise the low cost solutions with high savings in use.
2 High cost items with high savings in use need to have their value to the project demonstrated.
3 Low cost options having low savings in use should be reviewed regularly to ensure that options with higher savings are not available.
4 Options with high initial cost and low savings in use should be rejected for better alternatives.

As with new build these risk management appraisal techniques should also be used when evaluating life cycle cost for refurbishment projects.

12.3.4 Maintenance of mechanical and electrical systems

Traditional maintenance of mechanical and electrical (M&E) systems in the building is based on planned preventive maintenance (PPM) schedules. With the WLC model this approach is not the best since it does not target the unforeseen when an item of plant or equipment develops a fault in between scheduled maintenance intervals.

The preferred approach is to ensure that product manufacturers review their maintenance specifications to the ones based upon performance and condition monitoring (PCM) which are more cost-effective than PPM. See Appendix 3.

For both new build and refurbishments there are two questions to put to suppliers and manufacturers:

- How and when will the plant and equipment fail?
- Will parts be available within the life cycle being considered?

Opportunities for extending performance-based maintenance and continuing improvement and system reliability could eventually offer the possibility for manufacturers' warranties to be extended for the life cycle of the plant.

Clearly the traditional barriers between design and installation, operation and maintenance, must first be broken down and the need for appropriate disposal at the end of the cycle life to take its rightful place.

12.3.5 Supply chain management

The motor vehicle industry has used supply chain management techniques for many years. It involves a step change from the adversarial approach adopted by the building services industry towards suppliers and manufacturers to supply the product at the cheapest price and at the right time. This encourages manufacturers to only concern themselves with sales and leave to chance after-sales service and whether or not the product is the correct choice.

Supply chain management allows the building services engineer to encourage each manufacturer of each product to participate in the project design philosophy. This approach is at its most successful when suppliers and manufacturers of products required for the M&E services on a project actively participate in the design of the systems with the building services engineer, knowing what is required and offering innovative advice when appropriate.

This generates on the part of the product manufacturer a sense of ownership and responsibility for his contribution to the project, which is an essential outcome of supply chain management.

It is admitted that the element of competition is removed. However, it has been found that costs have not risen and the benefits far outway the adversarial approach that adversarial competition invariably brings.

12.4 Rethinking design and installation

Building services are particularly relevant to any attempt to predict the whole life costs of a building. They can account for 40–50% of the capital cost, they generate a significantly higher proportion of the operating costs, require regular maintenance and may need to be replaced at least once in the life of the building.

Reliable building services that last as long as they are meant to can offer significant savings to building operators by operating more effectively and gaining optimum performance from people and processes within the building as well as saving energy and Climate Change Levy in the process.

Porritt suggested four principles that engineers should try and grasp:

- Think function and services not profit.
- Adopt a full life cycle perspective.
- Think of breakthrough change rather than incremental improvement.
- Think of design for sustainability which is far more searching than design for the environment.

Clearly clients as well as the engineering organisations that are engaged by them need to be inspired to think in a sustainable way when considering a new building or refurbishment of an existing building.

Clients, building services engineers and product manufacturers all can profit from marketing the sustainable slogan since it is now in the public domain. It is now not uncommon to read in the technical press of innovative design in buildings and systems that clearly demonstrate low maintenance and low energy consumption.

Legislation both nationally and across Europe is forcing the issue and those who take up the challenges sooner rather than later will benefit.

12.5 Prefabrication

In the last 12 years some UK installation contractors of mechanical services in buildings have developed centralised workshops where a substantial part of the traditional installation work has been prefabricated off site. This approach has a number of benefits:

- Services are modulised and put together in controlled conditions that are not available on site.
- Electrical and mechanical tests can be certificated under controlled conditions.
- Partial commissioning can be undertaken at the prefabrication shop before delivery to site.
- Modules can be delivered to site just in time.
- Work on site is minimised as are potential problems.

It has been found that when this approach is adopted labour costs are reduced and completion deadlines are met and sometimes improved. It can also be seen that prefabrication will significantly assist in WLC projects. Five examples where off site modular prefabrication has been a success are given here.

- Multiple services in vertical service shafts
- Services in false ceilings

- Services in floor voids
- Services in toilet accommodation
- Plant rooms.

Coupling up the prefabricated modules on site has benefited from modern technology. Some product manufacturers have taken similar steps to installation contractors. Four examples will demonstrate this point:

- Some boiler manufacturers now offer prefabricated boiler plant for specific projects.
- Manufacturers of prefabricated bespoke manifolds for underfloor heating, radiators and wash hand basins.
- Up to six fan coil units can now be connected to a prefabricated manifold that reduces installation times and commissioning.
- A well known manufacturer of steam and condense plant and equipment has for some time offered modules prefabricated to a specific project's requirements.

Again it is clear that Whole Life Costing can benefit as it engages the manufacturer in taking specialist interest in the project and giving the opportunity for sharing break through innovation and design for sustainability.

12.6 An energy-saving product

Perhaps the most well known product that saves energy is the variable speed drive (VSD) for AC motors. There are countless examples now recorded where the cost of electrical power has been drastically reduced by appropriate selection of VSD control for pumps and fans. Savings of 60–70% are not uncommon providing a significantly short payback. The life of the product is also increased since it is not constantly working at full speed.

Depending on the application there are a number of other advantages in the use of VSDs

- Volume control dampers and regulating valves are not required to reduce the fan or pump flow rate to design value. This results in a speed reduction at all times when the system is operating at design conditions.
- The ideal soft start characteristics of VSDs leads to reduced maximum demand, less physical system stress, increased equipment life and improved reliability.
- Reduced fan speeds in particular leads to reduced noise levels.
- With the elimination of star delta starters and power factor correction and simplified monitoring via drive outputs, hidden installation costs are removed.

- Drive interrogation by serial communication, standard in many drives, means that every drive is self-monitoring reducing the need for sub-metering that Part L of the Building Regulations and EU legislation demand.

12.7   Energy-saving systems

The ultimate objective is to move away from the use of fossil fuel as a source of energy for heating and conditioning buildings. However, in the meantime serious reduction in the use of fossil fuel is the current requirement. This means reductions in the use of natural gas, oil and electricity that is derived from fossil fuel. The generation of green electricity is increasing both for the national grid and for local use.

There are six examples here of systems that reduce reliance on fossil fuel.

- Underfloor heating has reinvented itself in the light of modern well-insulated buildings. Forty years ago supplementary heating was required where underfloor heating was specified. One of its advantages is that it can operate at very low temperatures of around 50 °C. This means that it lends itself to the use of the condensing boiler or heat pump.
- Ground source heat pumps use ground temperature of around 12 °C to uprate to about 50 °C via a network of pipes horizontally or vertically placed in the ground.
- Thermal storage from solar panels during the summer months for use as space heating in the winter is now viable in well-insulated buildings.
- Combined heat and power is a serious contender increasingly for community heating. CHP, in conjunction with absorption cooling (known as tri-generation), can provide the reduction in CO_2 emissions called for in the Building Regulations for new buildings. This requires an improvement in energy efficiency and hence emissions from new plant *and* the provision of renewable energy or low carbon systems such as CHP. Clearly tri-generation using CHP and absorption cooling can achieve the objective set out in the Building Regulations.
- Solar thermal and microchip CHP are two realistic proposals for the existing housing stock in the UK. This sector is effectively more important than the new housing stock that is increasingly energy efficient due to current legislation. Refer to Section 3.9.
- Absorption cooling can reduce energy consumption when it is also used in tandem with waste heat. It can offer an environmentally and economically superior alternative to vapour compression refrigeration without the use of ozone-depleting refrigerants. The absorbant/refrigerant is known as the "working pair" and water/lithium bromide is one such pair. Another pair is water/ammonia. These refrigerants are benign.

The process uses (waste) heat from an external source to evaporate the refrigerant from the absorbant. It can then be condensed and reused. The absorption process works on the principle that the evaporation temperature of water is dependent upon atmospheric pressure. Therefore by creating a vacuum, the evaporation temperature can be lowered to a level suitable for use in refrigeration and air-conditioning. When used in conjunction with CHP it adds the facility of air-conditioning to heat and power generation by using some of the heat generated by the CHP unit. CHP systems are registered on the Government's Energy Technology List indicating that they are eligible for tax relief under the Enhanced Capital Allowances scheme (see Chapter 11). As part of CHP absorption chillers can also qualify.

12.8 Sustainable systems

A services system that saves energy during its life may not be a sustainable system. In Sections 12.1 and 12.2, the concept of a sustainable future for buildings and services is presented. Issues relating to the environment – in particular greenhouse gas emissions – and the reduction in consumption of fossil fuels only address climate change although this can lead to considerations on environmental impact and sustainability.

Buildings and services systems need to evolve so that less energy is consumed during their operating life on the one hand and on the other they need to be sustainable. That is to say their impact on the environment from sourcing the raw materials for production to final decommissioning and disposal at the end of their life cycle needs to be benign.

Services systems need to last for a specified life with prescribed performance condition monitoring (PCM) where appropriate, and at the end of the life to be removed, recycled and disposed of in a benign manner.

The costs in operating the mechanical and electrical systems in the building during the systems life must be determined at the design stage as whole life costing will need to be in place if systems are to be truly sustainable.

This requires the product manufacturers and the contractor who is responsible for the entire system to estimate the costs of maintenance and if necessary replacement over the system life. Product manufacturers will need to give assurances that the product, for example fans, pumps, boilers etc., will perform as defined in the PCM specification and the cost identified for the product life. The contractor responsible for the building services systems, with the help of the product manufacturers, will have to cost the operation of the systems over their life. This cost along with the capital cost of the systems represents the significant part of the whole life cost.

It is clear that this process of identifying operational life cost that *includes* the cost of energy to operate the products and systems is a relatively new concept in the building services industry. It is also a matter of managing the risk of unforeseen product or system failure.

Examples and case studies of annual energy consumption (AEC) and costs for a variety of buildings are included in the earlier chapters.

12.8.1 Using renewable energy

Clearly for a projected new or refurbished building and its services systems to be truly sustainable during the life cycle, energy must be sourced from renewables. This at present is a tall order and relies to an extent upon sourcing green electricity from the Generators and finding an alternative to fossil fuels. However, a system-by-system approach to first of all reduce annual energy demand (AED) to the lowest possible level in each of the services is feasible.

The AED for space heating, lighting, air conditioning, hot and cold water supply, security, fire detection, communications, vertical transportation will need to be analysed as suggested in Section 12.2.

The differences between AED and AEC can then be addressed.

Refer to Table 1.2 where seasonal efficiency = (AED)/(AEC). Clearly at this point the seasonal efficiency of each system then becomes the issue requiring attention.

12.8.2 Renewable energy

The scope of this book does not include an account of the various forms of renewable energy that are either currently in use or in the pipeline. However, Appendix 11 includes a plausible mix by 2020.

12.9 Chapter closure

This chapter has introduced the reader to a variety of issues facing the industry today. They relate to the way in which the industry conducts its business in the light of climate change and the move towards a sustainable future. The technology is available; it is attitudes that need to be re-thought to achieve different ways of doing business.

Appendix I

Standard Degree Day data

Degree Days can show for a given heating season how far outdoor temperatures are on average below the Base temperature (known also as the control temperature or the balance temperature) which is taken as 15.5 °C throughout this book. When outdoor temperature is above the Base temperature it should not be necessary to heat the building. This is because of the effect of indoor heat gains d caused from heat generated from equipment and lighting. Heat gains from occupants are not accounted for since this will have the effect of lowering the Base temperature and hence lowering the annual Degree Days and in consequence the Annual Energy Demand. Heat gains in this context are based upon fixed items of equipment that generate heat continuously during occupation of the building.

The higher the annual number of SDD the colder is the heating season. SDD are published by the Met. Office currently in Defra's *Energy Management* journal for every month of the year.

The annual 9-month SDD for the 20-year period up to May 1979 are given in Table 1.5, Chapter 1, for 17 locations in the UK.

The SDD given in this appendix is taken from the CIBSE Guide book A for the twenty year period ending in 1995 for 18 regions in the UK. If you compare the two tables they show that the climate is getting warmer.

Many organisations do not start up space heating plant now until 1 October thus reducing the heating season from 273 to 243 days. Account should be taken of this when using the SDD data. The month of May on the other hand is still considered as part of the heating season.

Notes relating to heating Degree Days.

i   The location area of North West Scotland has been included here making it the eighteenth region. It is not listed in Table 1.5, Chapter 1.

ii The annual SDD recorded for the 20-year period up to 1979 are generally higher than the 20-year average up to 1995, denoting a warmer climate for the later period of review.

iii It would be of interest to undertake a comparison between the 20-year average SDD to May 1979 listed in Table 1.5 with the 20-year average

SDD to May 1995 listed in this appendix in order to establish a trend. This comparison can be illustrated graphically by plotting on the same graph the location areas against the annual SDD for each location for the two periods of review.

Mean monthly and annual heating degree-day totals (Base temperature 15.5 °C) for 18 UK degree-day regions (1976–1995)

Degree-day region	Mean total degree-days (K · day)												
	Jan	Feb	Mar	Apr	May	Jun	Jul	Aug	Sep	Oct	Nov	Dec	Year
1 Thames Valley (Heathrow)	340	309	261	197	111	49	20	23	53	128	234	308	2033
2 South-eastern (Gatwick)	351	327	283	218	135	68	32	38	75	158	254	324	2255
3 Southern (Hurn)	338	312	279	222	135	70	37	42	77	157	246	311	2224
4 South-western (Plymouth)	286	270	249	198	120	58	23	26	52	123	200	253	1858
5 Severn Valley (Filton)	312	286	253	189	110	46	17	20	48	129	217	285	1835
6 Midland (Elmdon)	365	338	291	232	153	77	39	45	85	186	271	344	2425
7 W Pennines (Ringway)	360	328	292	220	136	73	34	42	81	170	259	331	2228
8 North-western (Carlisle)	370	329	309	237	159	89	45	54	101	182	271	342	2388
9 Borders (Boulmer)	364	328	312	259	197	112	58	60	102	186	270	335	2483
10 North-eastern (Leeming)	379	339	304	235	159	83	40	46	87	182	272	345	2370
11 E Pennines (Finningley)	371	339	294	228	150	79	39	45	82	174	266	342	2307
12 E Anglia (Honington)	371	338	294	228	143	74	35	37	70	158	264	342	2254
13 W Scotland (Abbotsinch)	380	336	317	240	159	93	54	64	107	206	286	358	2494
14 E Scotland (Leuchars)	390	339	320	253	185	104	57	65	113	204	290	362	2577
15 NE Scotland (Dyce)	394	345	331	264	194	116	62	72	122	216	295	365	2668
16 Wales (Aberporth)	328	310	289	231	156	89	44	44	77	156	234	294	2161
17 N Ireland (Aldergrove)	362	321	304	234	158	88	47	56	102	189	269	330	2360
18 NW Scotland (Stornoway)	336	296	332	260	207	124	85	88	135	214	254	330	2671

Heating degree-day and cooling degree-hour totals to various Base temperatures: London (Heathrow) (1976–1995)

Month	Heating degree-day total (K · day) for stated Base temperature (°C)								Cooling degree-hours (K · h) for stated Base temperature (°C)		
	10.0	12.0	14.0	15.5	16.0	18.0	18.5	20.0	5.0	12.0	18.0
January	165	224	286	332	348	410	425	472	1132	7	0
February	153	207	263	305	319	376	390	432	1070	14	0
March	104	158	217	263	278	340	355	402	2015	88	9
April	66	109	160	202	216	274	288	333	3031	402	38
May	19	43	78	112	124	176	190	233	5720	1509	252
June	2	9	27	48	57	97	109	145	7671	2857	641
July	0	2	7	17	22	48	57	87	9855	4682	1342
August	0	3	9	21	26	55	65	97	9416	4269	1071
September	4	13	31	53	62	106	119	159	7100	2361	285
October	26	52	91	129	143	201	216	262	4942	919	43
November	87	132	186	229	244	304	319	364	2507	188	0
December	138	193	254	300	316	378	393	440	1559	47	0
Year	764	1145	1609	2011	2155	2765	2926	3426	56018	17343	3681

Heating degree-day and cooling degree-hour totals to various Base temperatures: Manchester (Ringway) (1976–1995)

Month	Heating degree-day total (K · day) for stated Base temperature (°C)								Cooling degree-hours (K · h) for stated Base temperature (°C)		
	10.0	12.0	14.0	15.5	16.0	18.0	18.5	20.0	5.0	12.0	18.0
January	190	250	312	359	374	436	452	498	761	4	0
February	169	224	280	323	337	393	408	450	750	7	0
March	127	185	246	293	308	370	385	432	1415	33	3
April	83	131	186	229	243	302	317	362	2415	230	26
May	29	60	103	141	154	211	225	270	4775	967	120
June	5	18	45	74	85	133	146	187	6484	1878	323
July	1	4	16	34	42	81	93	131	8357	3243	639
August	1	6	21	41	50	93	105	145	7921	2864	474
September	8	23	52	83	95	148	162	205	5902	1405	84
October	38	73	122	164	179	240	255	301	4035	486	18
November	105	156	214	259	273	333	348	393	1919	70	0
December	165	223	285	331	347	409	424	471	1081	26	0
Year	921	1353	1882	2331	2487	3149	3320	3845	45815	11213	1687

Heating degree-day and cooling degree-hour totals to various Base temperatures: Edinburgh (Turnhouse) (1976–1995)

Month	Heating degree-day total (K · day) for stated Base temperature (°C)								Cooling degree-hours (K · h) for stated Base temperature (°C)		
	10.0	12.0	14.0	15.5	16.0	18.0	18.5	20.0	5.0	12.0	18.0
January	210	271	333	380	395	457	473	519	599	2	0
February	182	237	294	336	350	407	421	463	606	3	0
March	149	208	269	316	331	393	409	455	1138	15	0
April	102	155	212	256	271	330	345	390	1875	102	15
May	46	86	136	178	192	252	267	313	3762	505	41
June	10	29	63	97	109	162	176	218	5618	1275	134
July	2	9	29	53	63	111	124	166	7254	2272	235
August	5	14	35	61	71	120	134	177	6983	2101	196
September	16	37	73	109	123	179	194	238	5132	940	28
October	54	95	148	192	207	268	284	330	3453	301	5
November	126	180	238	283	298	358	373	418	1594	43	0
December	188	248	310	356	372	434	449	496	884	9	0
	1090	1569	2140	2617	2782	3471	3649	4183	38898	7568	654

Energy conservation measures

Preliminaries

There are three matters which must be investigated prior to a consideration of energy conservation measures.

- There is in place an active preventive maintenance programme for the services within the building. See Appendix 3.
- There is in place an active preventive maintenance programme for the building envelope.
- The occupants of the building are satisfied with the level of comfort provided by the services in the building.

Energy conservation measures – The building

- Roof insulation equivalent to 200 mm of glass fibre having a thermal conductivity of 0.035 W/m K
- Wall insulation by dry lining, cavity fill or external cladding to a thermal transmittance coefficient of 0.4 W/m² K or less.
- Single glazing changed to double glazing having at least a 12 mm cavity and the facility for controlled trickle ventilation in the window frame if the building is not air-conditioned.
- Sealing around doors and windows with air locks provided at entrance doors.
- Ensure automatic closure of fire safety doors on corridors off the lifts and stairwells.
- Ensure that the ventilation of the lift shaft occurs from outdoors to outdoors and that it is unaffected by strong wind.
- Toilet extract systems do not operate continuously but in response to operation of the artificial lighting system in the toilet or to door opening.

Air tightness of the building envelope

- Building Regulations Part 2L, 2002 for the first time introduced building envelope air tightness standards of $10 \, m^3/h \cdot m^2$ @ 50 Pa. Modern buildings can be constructed to achieve $5 \, m^3/h \cdot m^2$.
- The following example compares the annual energy loss for a model building having an envelope area of $5000 \, m^2$ with two levels of air leakage.

Example A2.1

$20 \, m^3/h \cdot m^2$ @ 50 Pa

Maximum natural air change 0.55 ac/h

Estimated energy loss through air leakage 111 725 kWh

$5 \, m^3/h \cdot m^2$ @ 50 Pa

Maximum natural air change 0.13 ac/h

Estimated energy loss through air leakage 27 932 kWh

The annual saving in energy is 83 793 kWh

This is equivalent to an annual saving of 15.92 tonnes of CO_2 from a natural gas-fired plant.

Building services – Space heating

- Check the time scheduling with the occupation times and the thermal response factor for the building. Optimum start/stop controls may be appropriate.
- If parts of the building have different occupation times consider local time scheduling via two or three port zone valves.
- Check thermostat settings on boiler plant and zones.
- The minimum temperature control for radiator systems is weather compensation on the boiler plant or weather compensated via a constant volume variable temperature control on a three port valve.
- There should be local temperature control available via thermostatic radiator valves or two port zone valves.
- Recent developments in boiler design have substantially increased the efficiency of heat conversion and at the same time reduced carbon dioxide and nitrous oxide emissions. Replacement should therefore be seriously considered if boiler plant is more than 10 years old with one of the new boilers operating in condensing mode.
- If there is only one heating boiler ask the question why.
- If there is more than one boiler, are sequence controls provided?
- Are time delays fitted to prevent the boiler plant from starting on a sudden temporary demand?
- Is the space heating boiler plant independent from the generation of hot water supply?

- If there is more than one building on the site is there a central boiler plant room? If so consider the provision of local plant when renewal is on the agenda to avoid losses in distribution mains.
- When pumps are replaced ensure that variable speed motors with the right speed controls are installed.
- Check the thermal insulation on distribution pipes.
- If windows are opened during cold weather find out why.

Building services – Hot water supply

- The generation of hot water supply should be independent of the space heating plant.
- Heat loss is sustained if the hot water is stored in a vessel prior to consumption – investigate the use of direct-fired instantaneous heaters.
- Check that hot water is only provided during the occupied period.
- Check the thermal insulation of storage vessels and distribution pipes.
- Check the storage or operating temperature, which should be a minimum of 60 °C.
- Ensure that secondary circulation is taken to a point local to the draw off point.
- Consider the use of spray taps on basins.

Catering services

- Provide local heating and ventilation to the kitchen area.
- Ensure that local temperature controls are provided and used by the catering staff.
- Ensure that local ventilation controls are provided and used by the catering staff.
- Consider providing local heating for domestic hot water.
- Install separate metering for the kitchen equipment.
- Ensure that kitchen equipment is modern and energy efficient.

Artificial lighting

- Check lighting levels in all areas with the recommended illuminance.
- Check the window cleaning schedule.
- Ensure that the luminaires are regularly cleaned and lamps replaced before they fail.
- Use energy efficient fluorescent tubes with electronic fittings.
- Ensure that GLS tungsten lamps are replaced if possible.
- Educate the occupants to switch the lights off.
- Install automatic lighting controls.

Small power equipment

- Ensure that equipment with standby mode does not have this facility disabled.
- Ensure that staff switch off equipment is not in use.
- Ensure that security staff switch off equipment not required when the building is unoccupied.
- Identify equipment that is not energy efficient.

Mechanical ventilation

- When was the air handling unit and distribution ductwork cleaned out?
- Check the time scheduling with the occupation times.
- Check the air flow rates at the supply and extract grilles and of the fans in the air handling unit.
- Check the temperature and where appropriate the humidity control settings.
- If the system does not provide for the space heating, must it operate at all times when the building is occupied?
- If the system operates on full fresh air only, consider the installation of a recuperator in the extract/supply air duct if this is not fitted.
- If there are substantial heat gains in the building is the sensible and latent heat being taken out from the extract air to heat the fresh air supply?
- When fans are replaced ensure that fans with variable speed control are fitted.

Preventive maintenance measures – Performance condition monitoring

Planned preventive maintenance plays an essential role in controlling the consumption of electricity and fossil fuel on a site. Poor maintenance and reliance upon corrective maintenance will lead to increases in energy use. The energy manager is well advised to ensure that a programme of planned preventive maintenance (PPM) is in place and strictly observed to the point where the manager builds into the programme his own checks and balances to verify the performance of the maintenance staff or contractors. This not only includes the presentation of signed certificates for each item of plant serviced for example, but also witnessing the maintenance work being undertaken even if it is only on a spot check basis. Manufacturers of key products should be encouraged in the system of supply chain management to review their maintenance specifications based upon performance condition monitoring (PCM).

Another essential element associated with the proper operation of the mechanical and electrical services within the building is the initial commissioning of the systems. The energy manager would be wise to ensure that the commissioning process had been properly undertaken and recorded for each of the services in the building.

The following list of preventive maintenance measures cannot be exhaustive and does not suggest frequency but hopefully gives an understanding of the breadth and depth of commitment required and all too frequently forgotten.

Space heating

- Plant room: should be clean and tidy and free from clutter. Check the fresh air intake if a combustion process takes place.
- Boiler plant: cleaning and adjustment to fuel burner, combustion test using a flue gas analyser identifying products of combustion and quantities. Air fuel ratio check. Thermostat settings, sequence controls.

Cleaning of boiler fire tubes, check on smokepipe and flueways. Thermal efficiency test. Check for leakage of flue gas and water.

- System: flushing out and use of inhibitor, feed and expansion tank – ball valve (partially submerged during system operation), open vent, overflow, cold feed, tank cover in place, thermal insulation *or* pressurisation unit and feed pump – operating pressure and temperature.
- Controls and control valves, TRVs: check actuator, operation and settings and time scheduling.
- Pumps and fans: check for noise, speed, power supply, pressure developed, belt tension and wear, leakage, lubrication.
- Distribution pipes: thermal insulation, valves, drain points, air eliminators.
- Space heating appliances: clean heat exchangers and filters in fan convectors/fan coil units/unit heaters/natural draught convectors. Check the fans as for pumps and fans.
- Radiators: clean out convection channels, ensure that they are painted with non-reflective paint, clean reflective panels behind the radiators, check the radiator valves for leakage and operation.
- Buildings usually over ten storeys: horizontal as well as vertical zoning with upper storeys having weather compensation set for more severe climate.

Central storage hot water supply

- Calorifier/indirect cylinder: check the heat exchanger for corrosion/scale, check the thermal insulation.
- Secondary pump: check noise, speed, power supply, pressure developed, leakage, corrosion and scale.
- Storage tanks: check contents, ball valve, open vent, overflow, cold feed, cover in place, insulation.
- Secondary flow and return: flushing out to remove scale, thermal insulation, valves and stopcocks, drain points.

Direct-fired hot water supply heaters

As for boilers and secondary pumps and secondary flow and return. In addition, since the heater is using raw water, the heat exchanger will require regular inspection.

These heaters are frequently fed from the rising main, and the following equipment should be subject to checking: strainer, pressure limiting valve, check valve, expansion valve, temperature/pressure relief valve, expansion vessel.

Plate heat exchangers for hot water supply

Due to their design, secondary water treatment may be necessary and regular descaling of the plates. An inhibitor should be considered for the primary water.

Mechanical ventilation

- Plant room: should be clean and free from clutter.
- Fans: as for space heating pumps.
- Filters: cleaned/replaced.
- Fresh air intake: debris cleared away.
- Air handling unit: check air heater batteries and cooling coils for leaks and dirt build up, recuperators for cleanliness, volume control dampers and linkages checked, casing checked for leaks.
- Controls and controllers: check actuators, operation and settings, thermostat, humidistat settings, time scheduling.
- Distribution ductwork: check for cleanliness, check fire dampers, volume control dampers, thermal insulation.
- Supply and extract grilles/diffusers: check directional louvre positions, flow rates, air temperatures.

Performance condition monitoring

As with the application of direct digital control software for systems, complex products like boilers, pumps, fans, fan coil units, air heater batteries, compressors, chillers, condensers etc. can have their performance monitored either centrally or locally so that condition monitoring can be undertaken. This process relies upon the manufacturer engaging with this concept and providing the software and monitoring points on the product.

It also engages the manufacturer in testing the product over its lifetime and identifying when specific maintenance or replacement of parts are needed. PCM is likely to be encouraged when a regime of supply chain management is in place. See Section 12.3.

Benchmarking

The current data relating to Benchmarking is given in the 2004 edition of the CIBSE Guide book F.

A series of booklets on energy efficiency was published between 1993 and 1995 by the Energy Efficiency Office of the Department of the Environment through BRECSU, the Building Research Energy Conservation Support Unit, to aid energy managers and facility managers responsible for one or more of thirteen different building types.

This was the next attempt at Benchmarking following benchmarking data included in the 1986 edition of Guide book B of the Chartered Institution of Building Services Engineers.

The benchmarking in the energy efficiency booklets goes by the terms Energy Consumption Yardsticks and Carbon Dioxide Yardsticks. The early booklets gave ECYs in GJ/m^2, Where $1\,GJ = 278\,kWh$.

These booklets have been freely available and are listed below.

Introduction to Energy
Efficiency in
Catering establishments
Entertainment buildings
Factories and warehouses
Further and higher education
Health care
Hotels
Museums, art galleries, libraries and churches
Offices
Post offices, building societies, banks and agencies
Prisons, emergency buildings and courts
Shops and stores
Schools
Sports and recreation centres

The Booklets each include information on energy management, the action plan, measures to achieve energy savings, energy use, energy consumption yardsticks and carbon dioxide yardsticks.

However, where possible it is recommended that the 2004 publication of CIBSE Guide book F is used for Benchmarking comparisons with a building's Performance Indicators.

Fossil and electric building benchmarks[1–22] (figures in shaded columns may be regarded as upper limits for new design)

Building type	Energy consumption benchmarks for existing buildings/(kW·h·m⁻²) per year (unless stated otherwise)				Basis of benchmark
	Good practice		Typical practice		
	Fossil fuels	Electricity	Fossil fuels	Electricity	
Catering[2]					
fast food restaurants	480	820	670	890	Gross floor area
public houses	1.5	0.8	3.5	1.8	(kW·h/m² per £1000 turnover)
restaurants (with bar)	1100	650	1250	730	Gross floor area
restaurants (in public houses)	2700	1300	3500	1500	(kW·h/cover[a])
Entertainment					
theatres	420	180	630	270	Gross floor area[b]
cinemas	515	135	620	160	Gross floor area[b]
social clubs	140	60	250	110	Gross floor area[b]
bingo clubs	440	190	540	230	Gross floor area[b]
Education (further and higher)					
catering, bar/restaurant	182	137	257	149	Gross floor area
catering, fast food	438	200	618	218	Gross floor area
lecture room, arts	100	67	120	76	Gross floor area
lecture room, science	110	113	132	129	Gross floor area
library, air conditioned	173	292	245	404	Gross floor area
library, naturally ventilated	115	46	161	64	Gross floor area
residential, halls of residence	240	85	290	100	Gross floor area
residential, self catering/flats	200	45	240	54	Gross floor area
science laboratory	110	155	132	175	Gross floor area
Education (schools)					
primary	113	22	164	32	Gross floor area
secondary	108	25	144	33	Gross floor area
secondary (with swimming pool)	142	29	187	36	Gross floor area
Hospitals[19]					
teaching and specialist	339	86	411	122	Heated floor area[d]
acute and maternity	422	74	510	108	Heated floor area[d]
cottage	443	55	492	78	Heated floor area[d]
long stay	401	48	518	72	Heated floor area[d]

(Continued)

Building type	Energy consumption benchmarks for existing buildings/(kW·h·m⁻²) per year (unless stated otherwise)				Basis of benchmark
	Good practice		Typical practice		
	Fossil fuels	Electricity	Fossil fuels	Electricity	
Hotels[16][e]					
holiday	260	80	400	140	Treated floor area
luxury	300	90	460	150	Treated floor area
small	240	80	360	120	Treated floor area
Industrial buildings					
post-1995; ≤5000 m²	96	–	–	–	Gross floor area
post-1995; >5000 m²	92	–	–	–	Gross floor area
pre-1995; ≤5000 m²	107	–	–	–	Gross floor area
pre-1995; >5000 m²	103	–	–	–	Gross floor area
Local authority buildings.					
car park (open)	–	–	–	1	Gross parking area
car park (enclosed)	–	–	–	15	Gross parking area
community centres	125	22	187	33	Agent's lettable area
day centres	203	51	349	68	Agent's lettable area
depots	283	37	311	39	Gross internal area
sheltered housing	314	46	432	68	Gross internal area
residential care homes	492	59	390	75	Gross internal area
temporary homeless units	408	48	467	71	Gross internal area
town hall (see also offices)	138	84	205	111	Gross internal area
Ministry of Defence (MoD) buildings.					
aircraft hangars (heated)	220	23	–	–	Treated floor area
junior mess	2.5	1.4	–	–	(kW·h per meal)
motor transport facilities	317	20	–	–	Treated floor area
multi-occupancy accommodation	225	29	–	–	Treated floor area
officers' mess	4.4	2.5	–	–	(kW·h per meal)
stores/warehouses (occupied)	187	34	–	–	Treated floor area
stores/warehouses (unoccupied)	54	3	–	–	Treated floor area
workshops	175	29	–	–	Treated floor area

Offices[10][f]					
air conditioned, standard	97	128	178	226	Treated floor area
air conditioned, prestige	114	234	210	358	Treated floor area
naturally ventilated, cellular	79	33	151	54	Treated floor area
naturally ventilated, open plan	79	54	151	85	Treated floor area
Primary health care (general practitioners' surgeries and dental practices)	174?	??	270?	??	Gross floor area
Public buildings					
ambulance stations	350	50	460	70	Treated floor area
churches	80	10	150	20	Treated floor area
courts (Magistrates)	125	31	194	45	Treated floor area
courts (County)	125	52	190	60	Treated floor area
courts (Crown)	139	68	182	74	Treated floor area
courts (combined Country/ Crown)	111	57	159	71	Treated floor area
fire stations	385	55	540	80	Treated floor area
libraries	113	32	210	46	Agent's lettable area
museums and art galleries	96	57	142	70	Gross internal area
police stations	295	45	410	60	Treated floor area
prisons	18861	3736	22034	4460	kW·h per prisoner[i]
prisons (high security)	18861	7071	22034	7509	kW·h per prisoner[i]
Residential and nursing homes	247	44	417	79	Gross floor area
Retail					
banks and building societies	63	71	98	101	Gross floor area
banks and building societies (all electric)	–	122	–	195	Gross floor area
book stores (all electric)	–	210	–	255	Sales floor area
catalogue stores	37	83	69	101	Sales floor area
catalogue stores (all electric)	–	100	–	133	Sales floor area
clothes shops	65	234	108	287	Sales floor area
clothes shops (all electric)	–	270	–	324	Sales floor area
department stores	194	237	248	294	Sales floor area
department stores (all electric)	–	209	–	259	Sales floor area
distribution warehouses	103	53	169	67	Sales floor area
distribution warehouses (all electric)	–	55	–	101	Sales floor area
DIY stores	149	127	192	160	Sales floor area
electrical goods rental	–	281	–	368	Sales floor area
electrical goods retail	–	172	–	230	Sales floor area
frozen food centres	–	858	–	1029	Sales floor area
high street agencies	150	55	230	75	Sales floor area
high street agencies (all electric)	–	90	–	160	Sales floor area
meat butchers (all electric)	–	475	–	577	Sales floor area
off licences (all electric)	–	475	–	562	Sales floor area

(Continued)

Building type	Energy consumption benchmarks for existing buildings/(kW·h·m⁻²) per year (unless stated otherwise)				Basis of benchmark
	Good practice		Typical practice		
	Fossil fuels	Electricity	Fossil fuels	Electricity	
supermarket (all electric)	–	1034	–	1155	Sales floor area
post offices	140	45	210	70	Sales floor area
post office (all electric)	–	80	–	140	Sales floor area
shoe shops (all electric)	–	197	–	279	Sales floor area
small food shops	80	400	100	500	Sales floor area
small food shops (all electric)	–	440	–	550	Sales floor area
supermarket	200	915	261	1026	Sales floor area
Sports and recreation:					
combined centre	264	96	598	152	Treated floor area
dry sports centre (local)	158	64	343	105	Treated floor area
fitness centre	201	127	449	194	Treated floor area
ice rink	100	167	217	255	Treated floor area
leisure pool centre	573	164	1321	258	Treated floor area
sports ground changing facility	141	93	216	164	Treated floor area
swimming pool (25 m) centre	573	152	1336	237	Treated floor area

Appendix 5

Monitoring equipment

There is a wealth of equipment available for hire or purchase that can monitor and record. The Building Services Research and Information Association (BSRIA) will hire out equipment for monitoring and recording purposes. Some of this equipment is listed here.

- Meters: digital electricity meters, digital gas meters, digital heat meters.
- Heat meter simulation: heat energy consumption in space heating sub-circuits can be determined in kWh or GJ by employing a portable ultrasonic flow meter and electronic thermometers clamped on to the circuit flow and return pipes and connected to a data logger.
- Air flow measurement: electronic vane anemometers, thermal vane anemometers, anemometer hoods (for measuring air flow from grilles), pito static tubes.
- Pressure measurement: micromanometers, Bourdon gauges, differential pressure gauge test sets, U tube differential pressure test sets, static pressure transducers.
- Duct leakage tests: portable test sets.
- Temperature measurement: digital thermometers, differential thermometers, infrared digital thermometers, infrared radiation thermometers, thermal imaging systems.
- Humidity measurement: wet and dry bulb whirling hygrometers, digital humidity indicators.
- Water flow measurement: computerised flow and differential pressure test sets, ultrasonic flow meters, micronics high temperature sensor set.
- Electrical measurement: digital induction ammeters, digital and analogue network testers, microprocessor controlled gaussmeter for measuring magnetic flux density, voltage condition analysers and multimeters, oscilloscope meters, power disturbance monitors, digital power and power factor indicators, electrical energy load analysers, optical reader for gas and elecricity meter readings, portable appliance insulation testers, mechanical/optical tachometers.

- Illumination inspection: illuminance meters calibrated in lux for illuminance measurement and Candela/m^2 for measurement of luminance.
- Combustion analysis: portable electronic combustion analysers, continuous sampling gas detectors, pocket gas monitors, hand-held combustion analysers for measuring oxygen, carbon monoxide, nitrogen oxide and flue gas temperature.
- Indoor air quality: carbon monoxide/carbon dioxide analysers, flammable gas monitors.
- Recording equipment: electromechanical chart recorders, microprocessor chart recorders, multichannel data printers, portable computers for storing data from chart recorders, loggers, combustion analysers etc., multichannel data loggers, mechanical/electronic thermohygrographs.

Appendix 6

Cost–benefit tables

These tables have been reproduced from the *CIBSE Guide*, Section B18 (1970) by permission of the Chartered Institution of Building Services Engineers.

Present value of a single sum

n (years)	Interest (= 100r) (%) 3	4	5	6	7	8	9	10	12	15	20
1	0.97087	0.96154	0.95238	0.94340	0.93458	0.92593	0.91743	0.90909	0.89286	0.86957	0.83333
2	0.94260	0.92456	0.90703	0.89000	0.87344	0.85734	0.84168	0.82645	0.79719	0.75614	0.69444
3	0.91514	0.88900	0.86384	0.83962	0.81630	0.79383	0.77218	0.75131	0.71178	0.65752	0.57870
4	0.88849	0.85480	0.82270	0.79209	0.76290	0.73503	0.70843	0.68301	0.63552	0.57175	0.48225
5	0.86261	0.82193	0.78353	0.74726	0.71299	0.68058	0.64993	0.62092	0.56743	0.49718	0.40188
6	0.83748	0.79031	0.74622	0.70496	0.66634	0.63017	0.59627	0.56447	0.50663	0.43233	0.33490
7	0.81309	0.75992	0.71068	0.66506	0.62275	0.58349	0.54703	0.51316	0.45235	0.37594	0.27908
8	0.78941	0.73069	0.67684	0.62741	0.58201	0.54027	0.50187	0.46651	0.40388	0.32690	0.23257
9	0.76642	0.70259	0.64461	0.59190	0.54393	0.50025	0.46043	0.42410	0.36061	0.28426	0.19381
10	0.74409	0.67556	0.61391	0.55839	0.50835	0.46319	0.42241	0.38554	0.32197	0.24718	0.16151
11	0.72242	0.64958	0.58468	0.52679	0.47509	0.42888	0.38753	0.35049	0.28748	0.21494	0.13459
12	0.70138	0.62460	0.55684	0.49697	0.44401	0.39711	0.35553	0.31863	0.25668	0.18691	0.11216
13	0.68095	0.60057	0.53032	0.46884	0.41496	0.36770	0.32618	0.28966	0.22917	0.16253	0.09346
14	0.66112	0.57748	0.50507	0.44230	0.38782	0.34046	0.29925	0.26333	0.20462	0.14133	0.07789
15	0.64186	0.55526	0.48102	0.41727	0.36245	0.31524	0.27454	0.23939	0.18270	0.12289	0.06491
16	0.62317	0.53391	0.45811	0.39365	0.33873	0.29189	0.25187	0.21763	0.16312	0.10686	0.05409
17	0.60502	0.51337	0.43630	0.37136	0.31657	0.27027	0.23107	0.19784	0.14564	0.09293	0.04507
18	0.58739	0.49363	0.41552	0.35034	0.29586	0.25025	0.21199	0.17986	0.13004	0.08081	0.03756
19	0.57029	0.47464	0.39573	0.33051	0.27651	0.23171	0.19449	0.16351	0.11611	0.07027	0.03130
20	0.55368	0.45639	0.37689	0.31180	0.25842	0.21455	0.17843	0.14864	0.10367	0.06110	0.02608
25	0.47761	0.37512	0.29530	0.23300	0.18425	0.14602	0.11597	0.09230	0.05882	0.03038	0.01048
30	0.41199	0.30832	0.23138	0.17411	0.13137	0.09938	0.07537	0.05731	0.03338	0.01510	0.00421
35	0.35538	0.25342	0.18129	0.13011	0.09366	0.06763	0.04899	0.03558	0.01894	0.00751	0.00169
40	0.30656	0.20829	0.14205	0.09722	0.06678	0.04603	0.03184	0.02209	0.01075	0.00373	0.00068
45	0.26444	0.17120	0.11130	0.07265	0.04761	0.03133	0.02069	0.01372	0.00610	0.00186	0.00027
50	0.22811	0.14071	0.08720	0.05429	0.03395	0.02132	0.01345	0.00852	0.00346	0.00092	0.00011
55	0.19677	0.11566	0.06833	0.04057	0.02420	0.01451	0.00874	0.00529	0.00196	0.00044	0.00004
60	0.16973	0.09506	0.05354	0.03031	0.01726	0.00988	0.00568	0.00328	0.00111	0.00023	0.00002

Note

The value of £1 in n years hence, when discounted at interest rate r per annum $= (1 + r)^{-n}$.

Terminal value of a single sum at compound interest

n (years)	Interest (= 100r) (%)										
	3	4	5	6	7	8	9	10	12	15	20
1	1.0300	1.0400	1.0500	1.0600	1.0700	1.0800	1.0900	1.1000	1.1200	1.1500	1.2000
2	1.0609	1.0816	1.1025	1.1236	1.1449	1.1664	1.1881	1.2100	1.2544	1.3225	1.4400
3	1.0927	1.1249	1.1576	1.1910	1.2250	1.2597	1.2950	1.3310	1.4049	1.5209	1.7280
4	1.1255	1.1699	1.2155	1.2625	1.3108	1.3605	1.4116	1.4641	1.5735	1.7490	2.0736
5	1.1593	1.2167	1.2763	1.3382	1.4026	1.4693	1.5386	1.6105	1.7623	2.0114	2.4883
6	1.1941	1.2653	1.3401	1.4185	1.5007	1.5869	1.6771	1.7716	1.9738	2.3131	2.9860
7	1.2299	1.3159	1.4071	1.5036	1.6058	1.7138	1.8280	1.9487	2.2107	2.6600	3.5832
8	1.2668	1.3686	1.4775	1.5938	1.7182	1.8509	1.9926	2.1436	2.4760	3.0590	4.2998
9	1.3048	1.4233	1.5513	1.6895	1.8385	1.9990	2.1719	2.3579	2.7731	3.5179	5.1598
10	1.3439	1.4802	1.6289	1.7908	1.9672	2.1589	2.3674	2.5937	3.1058	4.0456	6.1917
11	1.3842	1.5395	1.7103	1.8983	2.1049	2.3316	2.5804	2.8531	3.4785	4.6524	7.4301
12	1.4258	1.6010	1.7959	2.0122	2.2522	2.5182	2.8127	3.1384	3.8960	5.3502	8.9161
13	1.4685	1.6651	1.8856	2.1329	2.4098	2.7196	3.0658	3.4523	4.3635	6.1528	10.699
14	1.5126	1.7317	1.9799	2.2609	2.5785	2.9372	3.3417	3.7975	4.8871	7.0757	12.839
15	1.5580	1.8009	2.0789	2.3966	2.7590	3.1722	3.6425	4.1772	5.4736	8.1371	15.407
16	1.6047	1.8730	2.1829	2.5404	2.9522	3.4259	3.9703	4.5950	6.1304	9.3576	18.488
17	1.6528	1.9479	2.2920	2.6928	3.1588	3.7000	4.3276	5.0545	6.8660	10.761	22.186
18	1.7024	2.0258	2.4066	2.8543	3.3799	3.9960	4.7171	5.5599	7.6900	12.375	26.623
19	1.7535	2.1068	2.5269	3.0256	3.6165	4.3157	5.1417	6.1159	8.6128	14.232	31.948
20	1.8061	2.1911	2.6533	3.2071	3.8697	4.6610	5.6044	6.7275	9.6463	16.367	38.338
25	2.0938	2.6658	3.3864	4.2919	5.4274	6.8485	8.6231	10.835	17.000	32.919	95.396
30	2.4273	3.2434	4.3219	5.7435	7.6123	10.063	13.268	17.449	29.960	66.212	237.38
35	2.8139	3.9461	5.5160	7.6861	10.677	14.785	20.414	28.102	52.800	133.18	590.67
40	3.2620	4.8010	7.0400	10.286	14.974	21.725	31.409	45.259	93.051	267.86	1469.8
45	3.7816	5.8412	8.9850	13.765	21.002	31.920	48.327	72.890	163.99	538.77	3657.3
50	4.3839	7.1067	11.467	18.420	29.457	46.902	74.358	117.39	289.00	1083.7	9100.4
55	5.0821	8.6464	14.636	24.650	41.315	68.914	114.41	189.06	509.32	2179.7	22644
60	5.8916	10.519	18.679	32.988	57.946	101.26	176.03	304.50	897.59	4384.1	56346

Note
The amount to which £1 will increase in n years with interest rate r per annum $= (1 + r)^n$.

Present value of an annuity

Interest (= 100r) (%)

n (years)	3	4	5	6	7	8	9	10	12	15	20
1	0.9709	0.9615	0.9524	0.9434	0.9346	0.9259	0.9174	0.9091	0.8929	0.8696	0.8333
2	1.9135	1.8861	1.8594	1.8334	1.8080	1.7833	1.7591	1.7355	1.6901	1.6257	1.5278
3	2.8286	2.7751	2.7232	2.6730	2.6243	2.5771	2.5313	2.4869	2.4018	2.2832	2.1065
4	3.7171	3.6299	3.5460	3.4651	3.3872	3.3121	3.2397	3.1699	3.0373	2.8550	2.5887
5	4.5797	4.4518	4.3295	4.2124	4.1002	3.9927	3.8897	3.7908	3.6048	3.3522	2.9906
6	5.4172	5.2421	5.0757	4.9173	4.7665	4.6229	4.4859	4.3553	4.1114	3.7845	3.3255
7	6.2303	6.0021	5.7864	5.5824	5.3893	5.2064	5.0330	4.8684	4.5638	4.1604	3.6046
8	7.0197	6.7327	6.4632	6.2098	5.9713	5.7466	5.5348	5.3349	4.9676	4.4873	3.8372
9	7.7861	7.4353	7.1078	6.8017	6.5152	6.2469	5.9952	5.7590	5.3282	4.7716	4.0310
10	8.5302	8.1109	7.7217	7.3601	7.0236	6.7101	6.4177	6.1446	5.6502	5.0188	4.1925
11	9.2526	8.7605	8.3064	7.8869	7.4987	7.1390	6.8052	6.4951	5.9377	5.2337	4.3271
12	9.9540	9.3851	8.8633	8.3838	7.9427	7.5361	7.1607	6.8137	6.1944	5.4206	4.4392
13	10.6350	9.9856	9.3936	8.8527	8.3577	7.9038	7.4869	7.1034	6.4235	5.5831	4.5327
14	11.2961	10.5631	9.8986	9.2950	8.7455	8.2442	7.7862	7.3667	6.6282	5.7245	4.6106
15	11.9379	11.1184	10.3797	9.7122	9.1079	8.5595	8.0607	7.6061	6.8109	5.8474	4.6755
16	12.5611	11.6523	10.8378	10.1059	9.4466	8.8514	8.3126	7.8237	6.9740	5.9542	4.7296
17	13.1661	12.1637	11.2741	10.4773	9.7632	9.1216	8.5436	8.0216	7.1196	6.0472	4.7746
18	13.7535	12.6593	11.6896	10.8276	10.0591	9.3719	8.7556	8.2014	7.2497	6.1280	4.8122
19	14.3238	13.1339	12.0853	11.1581	10.3356	9.6036	8.9501	8.3649	7.3658	6.1982	4.8435
20	14.8775	13.5903	12.4622	11.4699	10.5940	9.8181	9.1285	8.5136	7.4694	6.2593	4.8696
25	17.4131	15.6221	14.0939	12.7834	11.6536	10.6748	9.8226	9.0770	7.8431	6.4641	4.9476
30	19.6004	17.2920	15.3725	13.7648	12.4090	11.2578	10.2737	9.4269	8.0552	6.5660	4.9789
35	21.4872	18.6646	16.3742	14.4982	12.9477	11.6546	10.5668	9.6442	8.1755	6.6166	4.9915
40	23.1148	19.7928	17.1591	15.0463	13.3317	11.9246	10.7574	9.7791	8.2438	6.6418	4.9966
45	24.5187	20.7200	17.7741	15.4558	13.6055	12.1084	10.8812	9.8628	8.2825	6.6543	4.9986
50	25.7298	21.4822	18.2559	15.7619	13.8007	12.2335	10.9617	9.9148	8.3045	6.6605	4.9995
55	26.6744	22.1086	18.6335	15.9905	13.9400	12.3186					
60	27.6756	22.6235	18.9293	16.1614	14.0392	12.3766					

Notes

The present value of £1 per annum for n years when discounted at interest rate r per annum = $[(1 - (1 + r)^{-n})/r]$.

The amount per annum to redeem a loan of £1 at the end of n years and provide interest on the outstanding balance at r per annum can be determined from the reciprocals of values in this table.

Levels and standards of artificial lighting

This table has been reproduced from the *CIBSE Guide*, Section A1 (1986) by permission of the Chartered Institution of Building Services Engineers.

Standard service illuminance (lx)	Characteristics of the activity/interior	Representative activities/interiors
50	Interiors visited rarely with visual tasks confined to movement and casual seeing without perception of detail.	Cable tunnels, indoor storage tanks, walkways.
100	Interiors visited occasionally with visual tasks confined to movement and casual seeing calling for only limited perception of detail.	Corridors, changing rooms, bulk stores.
150	Interiors visited occasionally with visual tasks requiring some perception of detail or involving some risk to people, plant or product.	Loading bays, medical stores, switchrooms.
200	Continuously occupied interiors, visual tasks not requiring any perception or detail.	Monitoring automatic processes in manufacture casting concrete, turbine halls.
300	Continuously occupied interiors, visual tasks moderately easy, i.e. large details >10 min arc and/or high contrast.	Packing goods, rough core making in foundries, rough sawing.
500	Visual tasks moderately difficult, i.e. details to be seen are of moderate size (5–10 min arc) and may be of low contrast. Also colour judgement may be required.	General offices, engine assembly, painting and spraying.

(Continued)

Standard service illuminance (lx)	Characteristics of the activity/interior	Representative activities/interiors
750	Visual tasks difficult, i.e. details to be seen are small (3–5 min arc) and of low contrast, also good colour judgements may be required.	Drawing offices, ceramic decoration, meat inspection.
1000	Visual tasks very difficult, i.e. details to be seen are very small (2–3 min arc) and can be of very low contrast. Also accurate colour adjustments may be required.	Electronic component assembly, gauge and tool rooms, retouching paintwork.
1500	Visual tasks extremely difficult, i.e. details to be seen extremely small (1–2 min arc) and of low contrast. Visual aids may be of advantage.	Inspection of graphic reproduction, hand tailoring, fine die sinking.
2000	Visual tasks exceptionally difficult, i.e. details to be seen exceptionally small (<1 min arc) with very low contrasts. Visual aids will be of advantage.	Assembly of minute mechanisms, finished fabric inspection.

Source organisations

CIBSE: Chartered Institution of Building Services Engineers, 222, Balham High Road, London, SW12 9BS

BSRIA: Building Services Research and Information Association, Old Bracknell Lane, Bracknell, Berkshire. RG12 7AH

HVCA: Heating and Ventilating Contractors Association, ESCA House, 34, Palace Court, Bayswater, London, W2 4JG

EEO: Energy Efficiency Office, address as for BRECSU. Note the previous title of the Energy Efficiency Office has been discontinued (April, 1996) and the work is now promoted under the name of Department of the Environments' Energy Efficiency Best Practice Programme

The Met. Office, Services and Business: Met. Office Fitzron Road Exeter EX1 3PB

ETSU: Energy Technology Support Unit, Harwell, Didcot, Oxfordshire, OX11 0RA

BRE: Building Research Establishment, Garston, Watford, WD2 7JR

BRECSU: Building Research Energy Conservation Support Unit, Enquiries Unit, Building Research Establishment, Garston, Watford, WD2 7JR

Building Regulations Approved Document Part L 2002 and 2006

Government White Paper on energy, March 2003

Appendix 9

Source journals

Building Services the CIBSE Journal
Defra's *Energy Management* Journal
BSEE: *Building Services Environmental Engineer* Journal
e.i.b.i: *Energy in Buildings and Industry* Journal

Energy saving initiatives

Government White Paper, March 2003
BREEAM: Building Research Establishment Environmental Assessment
  Method
EMAS: Ecological Management and Audit Scheme
  The European Community register for Local Authorities
TEWI: Total Equivalent Warming Impact Analysis
  Energy consumption of refrigeration plant over its working life
Action Energy
The Carbon Trust
EPBD European Energy Performance in Buildings Directive
Building Regulations 2005
SAP Standard Assessment Procedure
ISO 14001 A Voluntary International Standard for implementing,
  controlling and improving a company's Environmental Management
  System (EMS)

Renewable energy

The plausible mix of renewable energy systems by 2020 is likely to be:

Tidal
Wave
Geothermal
Photovoltaic
Solar thermal
Biomass
Offshore wind
Onshore wind
Hydro
Energy from waste
Nuclear reduced to about 5%
Fossil reduced to about 18%

By 2060 onshore wind and nuclear will be on their way to final extinction.

Source: Centre for Alternative Technology, 2004.

Word index

Index of examples and case studies